CONTROL

CONTROL

The Hidden Technology

ANDERS HANSSON

Linköping 2026

FSC
www.fsc.org
MIX
Papper från ansvarsfulla källor
Paper from responsible sources
FSC® C105338

The author generated this text and the cover in part with ChatGPT, OpenAI's large-scale language model.
The author generated interior illustrations using Yeri AI.

Publisher: BoD · Books on Demand, Östermalmstorg 1, 114 42 Stockholm, Sverige, bod@bod.se
Print: Libri Plureos GmbH, Friedensallee 273, 22763 Hamburg, Tyskland
ISBN: 978-91-8134-366-3

To Erik

Contents

Preface

This book is intended not only as a story about control, but also as a tool for education. Control systems now shape energy, transport, health care, buildings, media, manufacturing, defense, finance, and public infrastructure. For that reason, the book can be read by many different stakeholders in society: engineers who design automated systems, managers who must decide when automation is trustworthy, policy makers and regulators who need to understand technical promises and limits, teachers who want vivid examples, journalists and communicators who explain technology to the public, and citizens who live inside systems they rarely see. The chapters are written to make feedback, stability, sensing, actuation, autonomy, risk, and human responsibility understandable without requiring advanced mathematics. They can therefore support conversations between technical experts and non-specialists, especially where decisions about safety, comfort, efficiency, privacy, and accountability must be made together.

In university teaching, the book can be used as complementary reading wherever students are asked to connect technical systems with societal consequences: in control engineering, systems engineering, sustainability, technology and society, engineering ethics, artificial intelligence, robotics, and human–machine interaction. Its purpose is not to replace technical textbooks, but to give students a broader language for why control matters, where it succeeds, where it fails, and what responsibilities follow when machines are allowed to correct the world. This broader language is especially important for sustainability. Many of the challenges expressed in the United Nations Sustainable Development Goals depend on systems that sense, compare, decide, and act: clean energy must be balanced, cities must move and breathe efficiently, industry must reduce waste, health technologies must serve people safely, water and food systems must be monitored, and digital infrastructures must be governed responsibly. Control is therefore not only a technical discipline, but one of the practical means by which societies try to use resources wisely, reduce harm, increase resilience, and make collective promises measurable.

While this book is grounded in real science and historical records, the scenes involving dialogue and private interactions are narrative recreations. To illustrate control, the author has blended historical figures with composite and fictional characters. The events described are based on what could have happened, though they are presented as a dramatic interpretation rather than strict factual documentation.

Linköping, May 2026
Anders Hansson

Chapter 1

Measuring Time and Controlling the World

Better three hours too soon than a minute too late.

—William Shakespeare, *The Merry Wives of Windsor*, Act II, Scene ii

In the Athenian agora, long before a clock had hands and long before a second could be counted by the vibration of an atom, a slave named Thales, ordinary enough to have existed, carried a clay vessel through the morning crowd with both hands wrapped around its damp sides. He was not a philosopher, though his name invited jokes from men who were. He belonged to a household that lent him to the courts when speeches had to be timed, and on that morning he had been told to bring the clepsydra without spilling it, without cracking it, and without listening too carefully to citizens arguing about lawsuits as if the gods themselves were waiting for the verdict. Near the court, a magistrate named Philon, inspected the water clock with the practical suspicion of a man who knew that justice depended on a hole in a pot. "Is it clean?" he asked. Thales

tipped the vessel enough to show the small opening near the bottom. "Clean enough." "Clean enough is for sandals," Philon said. "Not for speeches." Beside them, Aristophanes of Byzantium, whom we may borrow from history even if the exact meeting is imagined, smiled at the seriousness of it. "You trust water with fairness?" he asked. Philon answered without looking up. "More than I trust men with unlimited time."

The Greek water clock was simple only to someone who did not have to use it. Water flowed from one vessel to another, or out through an orifice, and the falling level marked a portion of time. In the courts of Athens, the clepsydra helped decide how long a speaker could speak; in homes and sanctuaries, water clocks divided watches of the night; in later Hellenistic engineering, more elaborate forms used floats, pointers, and regulating devices to make the visible motion of time less dependent on the impatience of human eyes. The control problem was already present. Water does not flow at a perfectly constant rate merely because people wish to be fair. The pressure at the hole changes as the level falls. Sediment narrows passages. Clay swells. Temperature changes viscosity. A careless servant fills the vessel too high or too low. If the clock is to mean anything, the behavior of the water must be constrained, corrected, or at least made repeatable. Philon knew this in the language of civic anxiety rather than mathematics. "If the stream runs fast," he told Thales, "a poor man loses his defense. If it runs slow, a rich man buys the afternoon with words." Thales, who had spent his life listening to important men discover the importance of small things, said, "Then perhaps the water is the honest one, and the hole is the politician."

From the beginning, measuring time mattered because human beings needed to coordinate themselves with things that did not wait: daylight, prayer, irrigation, trade, ships, courts, meals, guards, markets, debts, labor, and the movements of the heavens. A community can live by sunrise and sunset when its demands are few and local, but the moment people must meet, exchange, judge, travel, build, or command at a distance, time becomes a public instrument. It is not enough to feel that the morning is passing; someone must say how much of it belongs to a speaker, a worker, a ritual, a journey, or a watchman on a wall. The clepsydra turned time into an object that could be carried into a room and pointed at. That transformation was moral as well as technical. It promised that two speakers might receive equal portions, that a guard's duty might end, that a priest might begin a rite at the proper hour, that a debtor could not stretch a day by pleading that the sun looked uncertain behind clouds. Control entered because measurement was never passive. To measure time was to build a device whose own behavior had to be disciplined before it could discipline society.

Centuries later, in an imagined workshop attached to a medieval monastery, Brother Martin woke before dawn not because he had measured the hour but because he feared he had failed to. The bell for matins was late, or early, or perhaps only his conscience was early, which in a monastery could amount to the same thing. In the tower above him, a new mechanical clock muttered in iron. Its weights pulled downward with the seriousness of falling stones; its gears translated that fall into rotation; its verge escapement stopped and released the gear train in a repeated rhythm that sounded less like flowing water than like argument. A visiting craftsman named Jean, had slept badly beside his tools and was already awake when Martin entered. "It hurries," Martin said. Jean rubbed his face. "Everything hurries when the weight is heavy and the teeth are new." "The abbot says prayer cannot depend on teeth." "Then the abbot should ask angels to ring the bell." Martin looked up toward the tower. The clock ticked, paused, ticked again, and in that pause lay the central invention: not motion, but interrupted motion, not power, but power restrained.

Mechanical clocks made the control problem visible in metal. A hanging weight or wound spring could provide energy, but energy alone does not make a clock; it makes a machine run away. The escapement is a governor of motion, a device that lets the gear train advance by small permitted amounts and then stops it again. Later pendulum clocks, transformed by the work of Christiaan Huygens in the seventeenth century, made this discipline far more precise by using the regular swing of a pendulum as the timekeeping element. Galileo had studied pendulum motion, Huygens made the pendulum clock practical, and generations of clockmakers learned that accuracy depended on controlling disturbances: friction in pivots, changes in temperature that altered pendulum length, variations in driving force, wear in teeth, dust, oil, and the violence of human winding. If the water clock asked how to make flow steady, the mechanical clock asked how to make stored energy obedient to rhythm. Jean tried to explain this to Brother Martin by lifting a small verge and letting it catch. "The weight wants to fall all at once," he said. "The clock says, not yet. Not yet. Not yet. That is time." Martin listened to the tick above them and said, "So the clock is a sermon against appetite." Jean considered this. "Against appetite, and against chaos, and against abbots who blame craftsmen for cloudy mornings."

The more accurate clocks became, the more the world reorganized around them. Monasteries needed hours of prayer, but cities needed market times, courts needed schedules, universities needed lectures, armies needed watches, and sailors needed longitude. At sea, the measurement of time became a matter of survival and empire. A navigator could estimate latitude from the heavens, but longitude required knowing the difference between local noon and the time at a reference meridian. That meant carrying home time across an ocean on a machine that would keep running through salt air, temperature changes, rolling decks, and months of vibration. John Harrison, the eighteenth-century English clockmaker, devoted his life to this problem with marine timekeepers that treated the ocean as an adversary to be controlled. In an imagined conversation in London, Harrison stands before a Board of Longitude examiner while a young sailor named Elias Reed, waits at the back of the room with a cap in his hands. "You ask whether a watch can find a ship," Harrison says. "I say a watch can remember where the ship came from." The examiner frowns. "A poetic answer is not a demonstration." Elias, whose brother was lost on a reef after a navigation error, mutters, "Nor is a dead crew." Harrison's chronometers were not merely beautiful mechanisms. They were control systems against the sea: balances, springs, temperature compensation, reduced friction, mechanisms designed so that the rate would not wander when the ship did.

By the nineteenth century, time measurement had moved from towers and ships into factories, railways, telegraph offices, and the nervous system of industrial life. A town could once set its clock by the sun and tolerate being several minutes different from the next town, but trains made local time dangerous. Two trains on one track cannot negotiate with the sun in each village. Telegraphy allowed time signals to travel faster than trains, and railways forced regions to standardize time so schedules could become enforceable promises rather than hopeful pamphlets. In a station office in 1883, on the day American railroads adopted standard time zones, a stationmaster named Clara Whitcomb resets a clock while an older farmer named Mr. Sloane watches with suspicion. "Noon has always been when the sun says noon," he tells her. Clara keeps her hand on the minute hand. "The sun is not dispatching the 2:17 to Chicago." "Chicago may keep its own sun." "Chicago is exactly the problem." The clock on the wall is no longer only a machine. It is a treaty among machines: locomotives, switches, telegraphs, watches, timetables, and human bodies that must arrive when the system expects them. Control now extends beyond the clock itself. The clock regulates work shifts, school bells, markets, mail, military orders, and the new discipline of punctuality. A person who once lived by tasks begins to live by appointments.

Electrical clocks deepened this transformation because they made time distributable, correctable, and centralizable. A master clock could send impulses to slave clocks throughout a building, a factory, a school, or a railway station, so that many faces showed one time. Synchronous electric clocks could lock their motion to the frequency of an alternating-current power grid, turning the stability of electrical generation into the stability of public time. Here the feedback loop became continental. Power systems had to regulate frequency because motors, clocks, and later countless devices depended on it; clocks revealed errors in the grid, and grid operators corrected generation to keep frequency near its nominal value. In an imagined control room in the 1930s, an engineer named Ruth Kaplan watches a frequency meter while her colleague Ben Alvarez listens to a wall of clocks click toward the hour. "If the load rises, the frequency sags," Ben says. Ruth nods. "Then the clocks lose time." "People complain about clocks before they complain about turbines." "Good," Ruth says. "Clocks are honest customers." The electrical clock shows how timekeeping and control can merge into infrastructure. The oscillator is no longer only in the clock case; it is spread through generators, governors, load changes, operators, meters, and the expectation that a city will wake, work, cook, and travel on a shared rhythm.

Quartz clocks made the oscillator smaller, cheaper, and more democratic. A quartz crystal vibrates at a stable frequency when driven electrically, and the clock counts those vibrations down into seconds. The essential control problem remains: the crystal must be cut, driven, protected, and interpreted; temperature changes must be reduced or compensated; circuits must divide high-frequency oscillation into usable ticks; batteries must provide power without letting the oscillator drift beyond usefulness. In a mid-twentieth-century laboratory, real engineers at Bell Labs, Seiko, and other companies helped move quartz timekeeping from specialized instruments into watches, radios, laboratories, and ordinary homes. Imagine a technician named May Chen holding a quartz wristwatch beside her father's mechanical pocket watch. Her father, a retired railway man, distrusts anything that makes no visible effort. "Where is the balance wheel?" he asks. "Inside, but not as you know it," May says. "Where is the tick?" "Too fast for you." He brings it to his ear and hears almost nothing. "A clock should sound like work." May smiles. "This one works by refusing to be dramatic." Quartz changed the social meaning of accuracy. Time precise enough for laboratories and navigation began to appear on kitchen walls, wrists, dashboards, and bedside tables. The more accurate clocks became, the less forgiving society became about lateness, synchronization, and delay.

Atomic clocks carried the story to a stranger and deeper level: time measured not by water, pendulums, wheels, grids, or crystals, but by the resonant frequency of atoms. In cesium atomic clocks, the definition of the second is tied to a specific transition frequency of the cesium-133 atom. The machine prepares atoms, exposes them to microwave radiation, observes how many make the desired transition, and uses feedback to tune the oscillator until it matches the atomic resonance. This is control in its most explicit form: compare an adjustable oscillator with a physical reference, detect error, correct the oscillator, repeat. In an imagined room at the National Physical Laboratory in Britain, Louis Essen, who built the first practical cesium atomic clock with Jack Parry in the 1950s, speaks with a graduate student named Amina Rahman. "So the atom is the clock?" Amina asks. Essen shakes his head. "The atom is the judge. The clock is the apparatus that keeps asking whether it is wrong." "And if it is wrong?" "We correct it." Amina looks at the racks of equipment, cables, dials, and vacuum hardware. "Then even the most perfect clock is an argument with error." Essen smiles. "Especially the most perfect clock." Atomic timekeeping does not remove control; it refines it until the feedback loop is aimed at nature's most reproducible rhythms.

The importance of measuring time has grown because modern life has become a coordination problem of extraordinary scale. Ancient courts needed fair speeches; monasteries needed prayer; sailors needed longitude; railways needed schedules; factories needed shifts; electrical grids needed frequency; radio and television needed synchronization; computers needed clocks to order operations; financial markets timestamp trades; telecommunications networks divide signals into precise intervals; satellites require exquisite timing to locate receivers on Earth. GPS is perhaps the clearest modern example. A receiver estimates position by comparing signals from satellites carrying atomic clocks;

an error of a tiny fraction of a second becomes a large error in distance. Time has become a form of space, money, law, memory, and power. In a conversation at a satellite control center, an engineer named Sofia Morales tells a visitor, "People think GPS tells them where they are. Really it tells them what time several satellites think it is, and the receiver turns disagreement into location." The visitor laughs. "That sounds like philosophy." Sofia points to a screen of orbital data. "It is engineering. Philosophy would let you be late."

The history of clocks therefore belongs at the beginning of a book about control because every clock is a promise that some unruly process has been made regular enough to trust. Water must flow predictably. Falling weights must be interrupted. Pendulums must be protected from friction and temperature. Springs and balances must resist disturbance. Electrical grids must hold frequency. Quartz crystals must be driven and counted. Atomic oscillators must be steered toward resonance. The user sees a dial, a hand, a number, or a timestamp; the engineer sees a fight against drift. Even the word accuracy hides a drama: a clock is always tempted away from truth by environment, wear, noise, and imperfect construction. Control is the art by which that temptation is noticed and resisted. Philon with his clepsydra, Brother Martin with his tower clock, Harrison with his marine chronometer, Clara with her railway clock, Ruth with her frequency meter, May with her quartz watch, and Amina with her atomic standard all face the same question in different centuries: what must be held steady so that human life can proceed?

At the end of the imagined day, Thales returns to the Athenian court to collect the empty water clock. The speeches are finished. One man is angry because time ended too soon; another is relieved because it ended at all. Philon rinses the vessel and holds it toward the light. "It is a small thing," he says. Thales looks at the clay, the hole, the last drops clinging to the inside. "Small things are safer to blame." Philon laughs, but Aristophanes of Byzantium, still lingering where no historical schedule requires him to be, says, "No. Small things are how large things learn discipline." The line may be too neat for real history, but it is true enough for control. Civilization did not become punctual all at once. It learned, drop by drop and tick by tick, that shared life depends on shared time, and shared time depends on devices that can correct, restrain, compare, and remember. Before the thermostat could hold a room near a chosen temperature, before a governor could steady an engine, before a rocket could balance on fire, human beings had already learned to ask a fragile machine to hold back chaos and say, with as much honesty as it could manage, now.

Chapter 2

The Thermostat and the Promise of Comfort

There is nothing like staying at home for real comfort.

—Jane Austen, *Emma*, Volume I, Chapter 10

The bell had not yet rung, and already the classroom was divided. Anna Carter sat near the tall window with her hands tucked beneath her arms. She had tried writing the date at the top of the page, but the letters came out stiff and narrow, as if the cold had entered the ink. Across the aisle, Thomas Bell had loosened his collar and was fanning his face with a folded sheet of paper. "If you open that window again," Anna said, "I shall lose the use of my fingers." Thomas glanced at the glass, where a gray Wisconsin morning pressed itself against the pane. "If I don't," he said, "I shall be cooked before Professor Johnson arrives." "You are not being cooked." "And you are not turning to ice." Anna gave him the look she had been practicing for future pupils: patient, disappointed, and meant to end a conversation. It did not work on Thomas, who grinned and reached for the latch. The window opened with a small wooden complaint. Cold air slid in low and fast, found Anna's ankles, and climbed. "Thomas." "There," he said. "A compromise." "A compromise would trouble both of us equally. This troubles me entirely."

Before Thomas could answer, Warren Johnson entered with a stack of papers under one arm and a piece of chalk in his hand. He was not a tall man in Anna's memory, though later she would distrust that memory because teachers often seemed larger while one was their student. He had a quick way of looking at a room, not simply seeing whether bodies were present, but taking the room's temperature in another sense: who was ready, who was absent in spirit, who had brought trouble in with his boots. That morning, he stopped before reaching the desk. The window was open. Anna was shivering. Thomas was flushed. Two students in the back row had not removed their coats. A boy near the stove had the sleepy, satisfied face of a cat. Johnson set down his papers. "Mr. Bell," he said, "are we studying ventilation today?" Thomas closed the window halfway. "No, sir." "Miss Carter, are we studying the polar regions?" "I hope not, sir." A few students laughed. Johnson smiled, but only briefly. He had come prepared to speak about expansion: how heat changes the dimensions of things, how a bridge, a rail, a wire, a strip of metal may move though no one sees it moving. It was a good lesson. He knew where he wanted it to go. But the room had already taken hold of it. He wrote one word on the board:

Heat.

The chalk made its dry sound. A coal wagon passed somewhere outside, iron-rimmed wheels grinding through the street. Inside, the class waited. "Heat," Johnson said, turning back to them, "does not merely make us comfortable or uncomfortable. It moves things. It changes their size. It changes their behavior. It changes what they are able to do." "It changes whether we can write," Anna murmured. Johnson heard her. So did half the class. Anna looked down at her page, ashamed of having spoken aloud. "Yes," Johnson said. "It does." The answer surprised her. She had expected correction, perhaps a reminder that a future teacher must not be ruled by inconvenience. Instead Johnson walked to her desk and looked at the thin, cramped line of writing at the top of the page. "May I?" She handed him the notebook. He held it up, not to mock her, but to show the room. "Here is our first demonstration." Anna went hot in the face. Thomas stopped smiling. "When Miss Carter came in," Johnson said, "she intended to write. She had paper, ink, attention, and purpose. The cold did not remove any of those. It merely made them harder to use." He returned the notebook gently. "That is often how the world defeats us. Not by forbidding a thing, but by making it just difficult enough that we do it badly."

No one laughed now. Johnson turned back to the board and drew a straight line. "A piece of metal expands when warmed. Slightly. Often too slightly for careless eyes. But the movement is real. If we can make use of that movement, we can make heat speak." Thomas leaned forward. "Speak, sir?" "In its own fashion." "Heat has a fashion?" "Everything has a fashion, Mr. Bell. The difficulty is learning how to listen." At that moment the door opened and Mr. Hale looked in. He was the man who kept the building alive. No one would have put it that way in the school catalogue. His name did not appear beside Johnson's in the list of instructors. But he knew the furnace, the drafts, the swollen doors, the windows that never quite latched, the rooms that chilled before noon and the rooms that became close and sour when the students filled them. He knew which teachers complained before the first recitation and which endured discomfort until irritation sharpened their voices. He knew the building the way a farmer knows a difficult animal. "You sent for me?" he asked. Johnson looked at the class. "I had not yet decided whether to send." "Someone did." From the back row, a boy raised one finger and then thought better of it. Mr. Hale sighed. "Too cold?" "Too warm," Thomas said.

"Too cold," Anna said, before she could stop herself. The room seemed to enjoy the contradiction. Mr. Hale stepped inside. He wore no coat, though his sleeves were rolled down and his hands were red from work. He looked first at the stove, then at the window, then at the students. He did not look like a man receiving information. He looked like a man confirming a suspicion. "If I open the draft," he said, "this side will complain by ten o'clock. If I leave it, that side complains now. If the window opens, the stove works harder and the hall turns cold. If the hall turns cold, Mrs. Whitcomb says her pupils cannot recite. If I warm Mrs. Whitcomb's room, the office becomes unbearable. Then the principal sends for me." He looked at Johnson. "So. What would you like?" It was not insolent. It was worse: it was honest. Johnson rested the chalk on the tray. "I would like the room to be fit for teaching." Mr. Hale laughed once, softly. "And I would like the building to have one opinion at a time." Thomas laughed. Anna did not. She was still holding her hands together under the desk. Johnson noticed. "Miss Carter cannot write," he said.

Mr. Hale turned to her, and his face changed a little. Complaints about rooms were common. A student unable to write was another matter. "Is that so?" Anna wished to disappear, but Johnson had made the truth public and therefore easier to own. "Only poorly," she said. Mr. Hale nodded. "I'll see what can be done." He left, and the door closed behind him. The class listened to his steps move down the corridor. Johnson did not immediately resume the lesson. He stood still, as if the interruption itself had become the lesson. "There," he said at last. "We have discovered a system." Thomas brightened. "The heating system?" "No. The complaining system." This time the laughter was warmer. Johnson walked down the aisle between the desks. "Miss Carter is cold. She tells us, or does not tell us. I notice, or fail to notice. Someone calls Mr. Hale. Mr. Hale believes us, or does not, and changes something elsewhere. Later, perhaps, the room improves. Or perhaps it becomes worse for Mr. Bell. Then Mr. Bell complains. Then Mr. Hale returns. Then we begin again." "It is not a very good system," Thomas said. "No." "It depends on everyone being annoyed." "A great many systems do." Anna smiled despite herself. Johnson returned to the front. "Suppose," he said, "the room could send word before Miss Carter's fingers stiffened."

"Send word to whom?" Thomas asked. "To the heat." "Rooms do not send word." "Not by nature." "Then by what?" Johnson picked up a thin strip of metal from the demonstration table. He had brought it for the lesson on expansion. In another version of the morning, he might have used it and moved on. But now the strip seemed to have acquired a different destiny. "By taking some small change seriously," he said. He held the strip between his fingers. "Warmth moves this. Not much. But enough, perhaps, if one is patient and clever. Cold moves it back. Imagine a device that waits in the room. It does not grow embarrassed. It does not decide whether a student is being troublesome. It does not forget to report. It simply feels the air and sends a signal." Anna tried to picture such a thing. A little messenger on the wall. A quiet servant of the room. Thomas pictured wires. "Would it ring a bell?" he asked. "It might." "Would Mr. Hale hate it?" "Almost certainly at first." The class laughed again, louder this time, and when the laughter settled the room felt less hostile, though no warmer. Downstairs, Mr. Hale adjusted what could be adjusted. He did it with the practical skepticism of a man who knows that every answer has a delay attached. Heat would come, but not instantly. By the time it arrived, the sun might reach the windows. The students might leave. The room might become too warm. People liked to imagine that he controlled the building. Mr. Hale knew better. He negotiated with it. By late morning, Johnson's classroom had improved, but imperfectly. Anna could write. Thomas had stopped fanning himself. The boy near the stove was asleep. Johnson saw him and tapped the desk. "Mr. Peters, has heat enlarged your understanding or only your comfort?" The boy woke to laughter.

Years later, Anna would remember that morning with more clarity than many formal lessons. She would not remember the exact equation, if there had been one. She would not remember every word Johnson said about metal. She would remember the way he had taken her discomfort seriously without making her seem weak. She would remember Mr. Hale in the doorway, burdened by the impossible task of satisfying a building full of bodies. She would remember Thomas saying, "The room knows first," though she might not be certain he had used those exact words. Memory polishes sentences. It also preserves truths. The room did know first. It knew before Anna complained. It knew before Johnson paused. It knew before Mr. Hale came up the stairs. The condition existed where the people were, while the power to change it lived somewhere else. Between those two places lay delay, hierarchy, embarrassment, and guesswork. Johnson did not invent discomfort. He did not invent the wish for warmth. He did not invent the fact that buildings are unfair. But he noticed that a room should not have to wait for a person to suffer visibly before it could ask to be changed. In 1883, Johnson patented an electric room thermostat. The name in the records is tidy and technical, as patent names tend to be. It does not contain Anna's cold fingers. It does not contain Thomas by the window. It does not contain Mr. Hale wiping his hands on a cloth while trying to satisfy five rooms with one furnace. A patent must describe a device. It cannot easily describe relief. But relief is the point. The thermostat gave the room a crude voice. It could not say, "A student is losing the thread of the lesson." It could not say, "The air near the window is different from the air near the stove." It could not say, "The girl in the front row is too proud to complain." It could say only, in its mechanical way, "The temperature has moved." That was enough to change daily life.

Before such devices, comfort often had to climb a ladder of permission. A child told a teacher. A teacher told an attendant. A worker told a foreman. A guest told a clerk. A patient told a nurse. The complaint traveled upward or outward, and at every step it could be dismissed, delayed, softened, forgotten, or judged. People who were bold, important, wealthy, or believed received comfort sooner than people who were quiet, junior, poor, or easy to ignore. Automatic temperature control did not make society equal. No machine on a wall could do that. But it removed one small humiliation from many rooms. It meant that warmth could arrive because the room had cooled, not because the right person had complained loudly enough. That is why the thermostat belongs in a story about human life and not only in a story about engineering. It helped turn buildings from places that had to be constantly pleaded with into places that could, sometimes, care for their occupants without being begged. Anna became a teacher. Let us allow her that future. On winter mornings in her own schoolhouse, she watched children enter with red hands and brave faces. She learned quickly that children do not all complain in the same way. Some announce discomfort as if it were a legal case. Some become foolish. Some become silent. Some misbehave because misbehavior is warmer than stillness. One boy, Samuel, came every day in a coat too thin for the season and insisted he was fine. Anna did not believe him. "Samuel," she said one morning, "change seats with Clara." "I am not cold, Miss Carter." "Then you will not mind proving it from the warmer side of the room." He moved. His handwriting improved before his pride did. There was no thermostat on Anna's wall at first. There was a stove, a draft, and her attention. But she had learned something from Johnson's room: comfort is not separate from learning. The mind does not float above the body like a balloon tied to a chair. It sits in the body. It borrows the body's ease. When the body is busy enduring, thought pays rent.

In time, buildings changed. Devices appeared on walls. A little dial, a small case, a hidden contact, later a digital display. People stopped thinking of these as inventions and began thinking of them as part of the wall, which is what successful inventions often become. They vanish into expectation. A hotel room should be warm. A hospital should not chill a patient in the night. A school should not require children to choose between attention and circulation. A greenhouse should not freeze because no one woke at two in the morning. A laboratory should not ruin an experiment because the weather shifted. A house should protect sleep. The thermostat did not create those desires. It made them easier to honor. Still, every gift casts a shadow. More than a century after Johnson's classroom, Ms. Rivera arrived early at a public school and found her room cold. The thermostat on the wall said otherwise. It sat behind a clear plastic cover with a small lock at the bottom. Ms. Rivera stood before it with her coffee in one hand and her lesson plans in the other. The display glowed with calm authority. "You are lying," she told it. The thermostat did not answer. Malik arrived first, hood up, backpack hanging from one shoulder. "Why is it always cold in here?" he asked. "Good morning, Malik." "Good morning. Why is it always cold in here?" Ms. Rivera laughed because she liked him and because the question was fair. "The system says it is not cold." He looked around the room. "The system should sit where I sit." By second period, three students were wearing coats. One girl had pulled her sleeves over her hands. Another said the room felt fine. A boy near the vent asked if he could move. Ms. Rivera began a discussion about metaphor and realized, halfway through a student's answer, that the room had become the metaphor. Everyone was speaking from where they sat.

At lunch, Mr. Chen from facilities came with a tablet. "I received a comfort request," he said. "A comfort request," Ms. Rivera repeated. "That is a beautiful phrase for mutiny." He smiled. He had the tired courtesy of someone blamed daily for invisible systems. "The sensor reads twenty degrees." "The students read colder." Mr. Chen looked at the sensor, then at the desks. "The sensor is on this wall." "The students are not." Malik, still in the room because he had nowhere better to spend lunch, pointed toward his seat. "Put the sensor over there." "If I could rebuild the system during lunch," Mr. Chen said, "I would begin with your desk." "So why not?" Mr. Chen leaned against a table. He did not speak down to Malik, which Ms. Rivera appreciated. "Because this classroom is part of a zone. The building groups rooms together. If I change this one, others change too. Some get too warm. Then those teachers call. Then I come back. Then everyone thinks I personally control winter." Malik considered this. "So the building thinks a bunch of rooms are one room." "More or less." "That's stupid." "Sometimes," Mr. Chen said. "Sometimes it is what the budget allowed." Ms. Rivera looked again at the thermostat. The old problem had not disappeared. It had become more polished. Instead of Mr. Hale walking from room to room, there was software, zoning, sensors, locked covers, energy targets, and work orders. The building listened better than Johnson's building had, but it still did not listen perfectly. It still believed some places more than others. "Can you make it better for tomorrow?" she asked. "I can try." That was an honest answer, old as furnaces. The next morning the room was better. Not perfect. Better. Malik entered, paused theatrically, and lowered his hood. "Tell the building I said thank you," he said. "Tell Mr. Chen yourself." "The building has people?" "All buildings have people," Ms. Rivera said. "That's the part they hide best." That, too, belongs in the story.

Automatic control can make care seem impersonal because it hides the people who designed, installed, maintained, paid for, adjusted, argued with, and depended on the system. When it works, the room is simply comfortable. No one thinks of Johnson. No one thinks of Mr. Hale. No one thinks of Anna's notebook or Malik's sleeves or Mr. Chen's tablet. Comfort becomes ordinary, and ordinary things are easy to undervalue. But ordinary comfort changed lives. It changed teaching by making classrooms less dependent on interruption. It changed hospitals by helping conditions remain steady through the night. It changed homes by allowing families to sleep without tending heat hour by hour. It changed workplaces, libraries, nurseries, laboratories, stores, theaters, and greenhouses. It gave people back small pieces of attention. That is not a small thing. Most of life is made from attention. A child learning to read. A nurse noticing a change in breathing. A machinist guiding a tool. A teacher hearing uncertainty in a student's voice. A parent sleeping deeply enough to be patient in the morning. If the environment constantly demands attention, it steals from all of these. The thermostat is a modest machine, but its promise is generous: let the background stay background. Of course, the background never fully obeys. Someone is still too cold. Someone is still too warm. The person with the locked cover has more power than the person under the vent. Energy is not free. Comfort in one room may carry a cost far away, in fuel, money, maintenance, or climate. A number on a wall can settle an argument too quickly. A sensor can be accurate and still not understand the room. But none of that erases the original kindness of the idea.

In Whitewater, after the students had gone, Johnson remained in the classroom. The heat had finally settled into something like comfort. The cruel joke was that the room often became pleasant after the people who needed it had left. Mr. Hale appeared in the doorway. "Better now?" Johnson looked at the empty desks. "Now, yes." Mr. Hale chuckled. "Rooms are often agreeable when no one's in them." Johnson picked up Anna's forgotten scrap of blotting paper from the floor and set it on the desk. "That is the trouble," he said. "What is?" "The room tells us what it needed after we needed it." Mr. Hale looked at him carefully. "Professor, rooms don't tell us anything." Johnson turned toward the window, where the evening had begun to darken the glass. "Then perhaps," he said, "we have been asking them the wrong way." He did not solve the problem in that moment. Inventions do not arrive whole because a man says something interesting in an empty room. There would be experiments, failures, drawings, wires, contacts, doubts, and money. There would be the hard work of making an idea behave in the world. But the human desire was already there, as plain as Anna's cramped handwriting. Let the room speak before the student suffers. Let the heat come before the lesson breaks. Let comfort depend a little less on complaint. After clocks taught machines to hold time steady, the thermostat brought the same promise into the inhabited room: not a machine taking command, but a room being given a small voice, so the people inside can return to their own.

Chapter 3

When Machines Learned Self-Control

Power tends to corrupt, and absolute power corrupts absolutely.
—Lord Acton, letter to Mandell Creighton, 5 April 1887

The first thing Samuel noticed was the sound. It was not one sound, exactly, but many sounds forced to live together: the knock of wood, the sigh of leather belts, the clatter of shafts overhead, the wet breath of steam, the cough of men clearing dust from their throats, the sharper voices of overseers trying to be heard above machinery that had no interest in human speech. The mill had a voice of its own, and when Samuel first entered it as a boy, he thought it sounded hungry. He was twelve, perhaps thirteen. No one at the door asked very carefully. His mother had tied a strip of cloth around his neck because the morning was cold, and because a mother will still try to make a child look cared for even when sending him into a place that will not care for him. She had smoothed his hair with wet fingers. "Do as you're told," she said. "And keep away from anything that moves." This was impossible advice. Everything moved. A man named Mr. Arkwright met him just inside the weaving room. He was not the famous

Arkwright; he merely carried the name as if it entitled him to impatience. His waistcoat strained at the buttons. His watch chain flashed when he turned. "You can crawl?" he asked.

Samuel looked at him, unsure whether this was a trick. "Yes, sir." "You can listen?" "Yes, sir." "Then you can learn. If a thread breaks, you call. If waste falls where it shouldn't, you clear it when told. If a belt slips, you do not touch it. If the engine changes its temper, you step back." "Changes its temper?" Mr. Arkwright looked toward the end of the room, where a shaft turned with a steadiness that seemed almost calm until one watched it too long. "You'll know." Samuel did not know. Not that first morning. He knew only that the mill was full of things that turned, pulled, struck, opened, closed, tightened, loosened, and swallowed. He knew that men had built the place, but the place did not feel built so much as awakened. It had a pulse. It had moods. When the machinery ran smoothly, people moved inside it with practiced indifference. When something changed, every adult in the room seemed to hear it before Samuel did. A woman at a loom lifted her head. A mechanic near the wall stopped wiping his hands. Mr. Arkwright said, very softly for a loud man, "There." Only then did Samuel hear it: the shaft was running faster. Not much faster. That was the frightening part. Nothing had broken. No wheel had flown apart. No belt had snapped. But the room had tightened. The machines had gained a little speed, and that little was enough to alter the faces of the people who knew what little could become. "Tom!" Mr. Arkwright shouted.

A broad-shouldered engineer appeared from below, blackened at the cuffs and jaw. His name was Thomas Price, though most called him Tom because machinery did not leave time for full names. He ran toward the engine room, and Samuel, forgetting his mother's instruction, followed far enough to see through the doorway. The engine was not beautiful in the way a gentleman's instrument might be beautiful. It was too large for that, too hot, too necessary. Rods moved with the confidence of limbs. A beam rose and fell. Steam entered, pushed, escaped. The whole thing seemed to breathe for the building. Tom reached for a lever. "Not so hard," another man warned. "I know my hand," Tom said. He eased the engine back. The room's speed settled. The invisible fear loosened. People returned to their work, though not quite to their former ease. Samuel stayed by the doorway until Tom noticed him. "You belong in there," Tom said, pointing back to the weaving room. "Mr. Arkwright said the engine changes its temper." Tom wiped his hands on a rag. "Mr. Arkwright says many things." "Does it?" "Does it what?" "Have a temper?" Tom looked at the engine, then at the boy. "No. Men have tempers. Machines have causes."

Samuel considered this and decided it was not entirely true. The engine did not have a temper perhaps, but it caused them in others. That evening, when Samuel returned home, his mother asked what work he had done. He told her about the looms, the heat, the noise, the lint in the air. Then he told her about the engine speeding up. "Did anyone get hurt?" she asked. "No." "Then thank God." "Tom stopped it." His mother stirred the pot over the fire. "Then thank Tom as well." For many years, engines depended on men like Tom. They depended on hands that knew the right amount, eyes that could read motion, ears that could detect a change in rhythm before it became disaster. This is easy to romanticize from a distance. We like the image of the skilled engineer standing beside the great machine, mastering it by experience. There was skill there, and pride. But there was also exhaustion. A man cannot watch perfectly forever. A man cannot be in every room that his engine affects. A man cannot always correct quickly enough, gently enough, early enough. The mill did not need heroism once a week. It needed steadiness every minute.

A few weeks after Samuel began work, a visitor came to the mill. He arrived with Mr. Arkwright and a younger man carrying a leather case. The visitor had silver at his temples and a face that looked tired rather than old. He was dressed better than the mechanics, but he did not look at the machinery like a gentleman being shown a curiosity. He looked at it as if it might answer questions. "Mr. Watt," Arkwright said, loudly enough for nearby workers to hear, "you see we run your engine hard here." James Watt did not smile at the flattery. "So I hear." Samuel heard the name pass through the room. Watt. Even boys knew some names before they knew why they mattered. Watt was spoken of as a man who had changed steam from a brute into a servant, though in Samuel's experience servants did not shake buildings. Tom Price came up from the engine room, wiping his hands. "Mr. Watt," he said. "Mr. Price." They spoke like men who had met before and disagreed usefully. "Still chasing speed?" Watt asked. Tom nodded toward the shafting. "Still being chased by it." Arkwright frowned. "The work requires regular motion. The looms complain if the engine wanders." "Looms are not the only ones," Tom said.

Watt looked up at the turning shaft. "How often?" "Often enough." "Load changes?" "Belts, looms, men starting and stopping things they should not. Steam pressure not always what it ought to be. Coal not always what it was promised to be. The usual family of sins." Watt's eyes moved from the shaft to the engine and back again. Samuel had crept closer, carrying a basket of waste he had no urgent need to carry. "And you correct by hand?" Watt asked. Tom held up his right hand. "With this miracle of modern science." Watt did smile then. "It is a good hand," he said. "But it must sleep." Tom looked at him. "Engines do not." The younger man with the leather case opened it on a table. Inside were parts wrapped in cloth: small rods, a spindle, a collar, jointed arms, and two metal balls that caught Samuel's eye immediately. They were not large, but they seemed important because they were the only things in the case that looked as if they wanted to move. "What are those?" Samuel whispered. The younger man glanced at him. "Trouble, if Mr. Price is stubborn." "Help, if Mr. Price is tired," Watt said. Tom crossed his arms. "I've seen governors on millstones." "Then you have seen the idea," Watt replied. "Not necessarily the answer." Arkwright leaned in. "It will keep the speed regular?" "It will try," Watt said. Samuel liked that answer. Adults rarely admitted when a machine would only try.

They fitted the device where it could be driven by the engine. Samuel watched from the edge of permission. Two balls hung from arms attached to a vertical spindle. When the spindle turned slowly, the balls hung low, like tired birds. When it turned faster, they lifted outward. Their rising moved a linkage. The linkage touched the throttle. The throttle governed steam. No one explained all this to Samuel in a single lecture. He learned it by watching faces. When the balls hung low, Tom watched the engine. When the balls rose, Tom watched Watt. When the linkage moved, Arkwright stopped breathing through his mouth. "So," Tom said, "the faster it runs, the more it chokes itself." "Not chokes," Watt said. "Restrains." "A polite choking." "If you like." "And if the speed falls?" "The balls fall. The valve opens. More steam enters." Tom looked unconvinced, but not uninterested. "It corrects both ways." "As a good man should," Watt said. "Few do." "Then perhaps machines may improve us." Tom snorted. "Or imitate our worst habits with better timing." They started the engine. At first nothing seemed different. The beam moved. The shaft turned. Belts carried motion into the long room. The looms resumed their clatter. Samuel stood where he could see both the governor and the people watching it. The balls spun in a blur, lifting until they found a place in the air and stayed there, neither drooping nor flying apart. They looked alive, though not like an animal. More like a thought held in balance. Then a loom was engaged at the far end of the room. The load changed. Samuel did not know the word load yet, but he saw the consequence. The engine slowed slightly. The balls dipped. The linkage moved. Steam entered. The speed recovered. No shouting. No running. No Tom's hand on the lever. The machine had felt the change in its own motion and answered before the room had time to become frightened. A woman at one of the looms looked up, puzzled by the absence of drama. "Did it hold?" Arkwright asked. Tom did not answer immediately. "Again," he said.

They tried again. A belt was shifted. A machine was brought on. Another was eased off. Each time the balls moved. Each time the engine corrected. Not perfectly. Not magically. But calmly enough that the room seemed to gain a new kind of patience. Samuel realized he had been waiting for Tom to run. Tom did not run. He stood beside Watt with his arms folded, watching the spinning balls as if they had insulted and relieved him at the same time. "Well?" Watt asked. Tom wiped his jaw with the back of his hand. "It is a strange thing to see a machine mind itself." "It does not mind itself." "No?" "It is arranged so that its motion corrects its motion." Tom looked at him. "That is what I said, only colder." Watt accepted this.

By noon, everyone in the mill knew about the new device. No one used the same words for it. Arkwright called it an improvement. Watt called it a governor. Tom called it the spinning magistrate. Samuel called it the two iron birds. At dinner, the workers argued over what it meant. "If the engine watches itself," said a weaver named Mary Ellis, "will they need Tom less?" Tom, who was eating bread beside the wall, answered without looking up. "They will need me differently." "That is what men say when they are about to be replaced."

"No," Tom said. "When a machine replaces you, the owner says it. When the machine changes your work, you say it first so it sounds less frightening." Mary laughed, but not unkindly. Samuel sat with his bread untouched. "Does it think?" he asked. "The governor?" Mary said. He nodded. Tom tore his bread in half. "No." "But it knows when the engine is too fast." "It moves when the engine is too fast." "What's the difference?" Tom looked toward the engine room. The question deserved more than dismissal. "Knowing is what your mother does when she sees your face and asks what you've broken. Moving is what a stone does when dropped. The governor is closer to the stone." Samuel thought about this. "But a stone doesn't help." "Not usually." Mary leaned toward him. "Do not let Mr. Price take all the wonder out of it. A thing need not think to change the world."

That afternoon, the mill felt different. Not quieter; mills did not become quiet. Not safer in any complete sense; belts still moved, fingers were still fragile, lungs still filled with dust, wages were still thin, hours still long. But one particular fear had been reduced. The engine's speed no longer depended entirely on a man's continuous attention. The correction began inside the motion itself. This changed the workers' bodies before it changed their opinions. Tom stopped holding himself quite so ready to spring. The weavers looked up less often. Arkwright complained about production with renewed energy, which was a sign that one of his older worries had been partly satisfied. Samuel, who had been afraid of the engine as one fears an animal, began to fear it as one fears a law: still powerful, still capable of harm, but less moody. The governor did not make the mill kind. That must be said plainly. It did not shorten Samuel's day. It did not soften Arkwright's voice. It did not clean the air or raise wages. It did not make the machines less hungry for thread, coal, oil, and human time. A control device can stabilize a cruel system as easily as a generous one. Steadiness is not the same as justice. But steadiness matters. A loom that runs too fast breaks threads. A mill that stops and starts wastes labor. An engine that races risks itself and everyone around it. A worker who must listen for danger every second grows old in a particular way. Regular motion meant fewer sudden panics, fewer spoiled pieces, fewer frantic corrections. It meant that some part of the work no longer depended on catching trouble after it had already announced itself. For

Tom, the governor was both relief and insult.

Late that evening, after the workers had gone and the engine slowed toward rest, he remained with Watt in the engine room. The governor's balls hung lower now, turning lazily. "You don't like it," Watt said. Tom put away a tool. "I like it enough." "That is not the same thing." "No." Watt waited. Tom looked at the governor. "When I was apprenticed, my master said an engine speaks through its irregularities. You learn the knock, the hiss, the little hurry in the beam. You learn when to touch and when to leave be. A poor man hears noise. A good man hears intention." "And now?" "Now that thing hears some of it first." Watt's face softened. "Some." "Enough to make a man wonder what remains his." "Judgment," Watt said. "Repair. Doubt. Blame, when it fails." Tom laughed. "Then I am secure." Watt placed a hand lightly on the rail near the engine. "I do not think machines remove men from responsibility. They change the shape of it." "Owners do not always hear that part." "No," Watt said. "They often hear only profit." For a while neither spoke. Steam ticked in a cooling pipe. "It will fail someday," Tom said. "Yes." "Stick. Wear. Be badly fitted. Be trusted by a fool." "Yes." "And when it fails, it may fail faster than a man can understand." Watt nodded. "That is why a governor must not become an excuse to stop listening." Tom looked at him. "Will you tell Arkwright that?" "He will hear the first half." "That it governs." "Yes." "Not the second." "That it must be watched." Tom smiled without pleasure. "Then we have invented a very human machine."

Samuel did not hear this conversation. He had gone home, where his mother was mending a sleeve by firelight. But he carried the day with him. "It has iron birds," he told her. "What does?" "The engine. They spin. When it runs too fast, they fly outward and pull it back." His mother looked up from the sleeve. "Like reins?" "Maybe." "Then it is a horse after all." "Tom says machines have causes, not tempers." "Tom is a man. Men say things like that so they needn't admit they are afraid." Samuel sat beside the fire. He wanted to explain what he had seen, but the words were difficult. "Before, if the engine changed, Tom had to catch it. Today it caught itself." His mother stopped sewing. "That sounds useful." "It was." "Does it make the work safer?" Samuel thought of the belts, the looms, the boy near the carding machine with a scar across two fingers.

"Some of it." She nodded, accepting the honest size of the answer. "Then be grateful for some." Years later, Samuel would remember the governor not as a technical device but as a change in posture. Before it, certain adults in the mill stood as if waiting to be struck by news. After it, they stood a little differently. Not relaxed. No one in the mill was relaxed. But less haunted by one particular kind of surprise.

That is one of the quiet ways control engineering enters human life. It does not always arrive as a miracle. Sometimes it arrives as a reduction in flinching. A machine that once required constant correction begins to correct itself. A person who once had to watch every second can look away for a moment. A room stays warm. A shaft keeps speed. A patient sleeps while a device watches. An aircraft holds its course while a pilot studies weather ahead. The change is not that humans become unnecessary. The change is that attention is redistributed. And attention is life. The centrifugal governor became one of the great symbols of automatic control because it made a simple bargain visible. If the engine ran too fast, the spinning balls rose and reduced the steam. If it ran too slowly, they fell and allowed more. The machine's own motion became the source of correction. It was not intelligence, but it looked like a primitive form of self-command. People noticed. Some saw elegance. Some saw profit. Some saw danger. Some saw a future in which machines would need fewer hands. Some saw a future in which machines could be trusted with more power. All of them were partly right. The governor did not merely regulate engines. It changed the imagination of machines. A machine no longer had to be only driven. It could also be restrained by its own behavior. It could contain a small internal argument between motion and limit. This idea would travel far beyond mills. It would reappear in engines, turbines, clocks, vehicles, factories, power stations, aircraft, and devices no one in Samuel's mill could have named. But in the mill, before it became a principle, it was a pair of spinning balls that let people breathe a little easier.

One winter afternoon, several months after the governor was installed, the mill lost a belt. It snapped not far from Samuel, cracking through the air like a whip. He dropped flat because he had learned that fast things deserve the floor. The loose end struck a wooden guard and tore it away. A woman screamed.

The shaft jerked free of part of its burden, and for one terrible moment the engine began to hurry. Samuel looked toward the engine room. Tom was already moving. So was the governor. The balls rose sharply. The linkage pulled. Steam was checked. The engine did not race as it might have. Tom reached the lever a breath later and brought the system down further. Men shouted. The broken belt slapped once more and fell. Dust drifted in the sudden partial quiet. No one spoke for a moment. Then Mary Ellis said, with a shaking voice, "I dislike excitement before tea." Laughter came, thin but real. Tom inspected the damage. Arkwright complained about lost time before asking whether anyone was hurt, which told Samuel something about the order of values in the mill. No one was badly hurt. That became the fact people repeated. No one was badly hurt.

Samuel watched Tom examine the governor afterward. He checked the joints, the collar, the linkage, the valve. His hands moved carefully, almost respectfully. "It helped," Samuel said. Tom did not look up. "Yes." "You helped too." "That is the arrangement I prefer." "Man and machine?" "Man, machine, and suspicion." Samuel smiled. "Suspicion?" Tom tightened a nut. "Trust a device enough to let it help. Distrust it enough to keep it honest." That sentence stayed with Samuel longer than anything Watt had said. It was a worker's philosophy of automation, born not from fear of machines or worship of them, but from standing close enough to know what they could do. Trust it enough to let it help. Distrust it enough to keep it honest. By the time Samuel was grown, governors were no longer strange to him. He had seen them on engines in mills and workshops, had watched them spin in summer heat and winter gloom, had heard men curse them when badly adjusted and praise them when disaster did not occur. He learned that no device ends the need for care. A governor could hunt, sticking the engine into a restless rising and falling. It could be sluggish. It could be tampered with by men who wanted more speed than wisdom allowed. It could be ignored because it usually worked. The worst failures often begin with usually. Usually the governor holds. Usually the belt lasts. Usually the operator notices. Usually the machine sounds like that. Control makes life safer by making some things usual. It endangers life when usual becomes invisible.

Samuel learned this not from books, but from days. He learned it in the way workers learn: through near misses, burns, jokes after fear, funerals when jokes failed, and the ordinary pride of keeping something dangerous in service without letting it kill anyone. He came to respect the governor not because it made the engine harmless, but because it made one kind of harm less likely. That is often the honest measure of engineering. Not salvation. Less harm. Not perfect freedom. Less fear. Not a machine that thinks. A machine that corrects soon enough for people to continue living beside it. At the end of his life, if anyone had asked Samuel what the governor meant, he might not have spoken of Watt or patents or the history of steam. He might have described the first day he saw the balls rise. He might have described Tom Price standing still when once he would have run. He might have described Anna-like children in other places, though he never knew Anna, and how so many inventions begin with someone trying to do ordinary work under conditions that make ordinary work difficult. He might have said this: before the governor, the engine had to be watched like a temper. After the governor, it had something like manners. Not goodness. Manners. A mannered machine can still serve a harsh master. A steady engine can power exhausting labor. A regulated factory can produce both wealth and misery. Control does not decide what a system is for. It only helps the system pursue its purpose more reliably. That is why control engineering is morally interesting. It can protect people from chaos, but it can also make an unjust order more efficient.

The governor therefore gives us the third lesson of this book. The clock showed a device learning to hold time steady enough for shared life; the thermostat showed a room learning to speak before discomfort became complaint. The governor shows a machine learning to restrain itself before power becomes danger. In each case, the promise is humane: fewer interruptions, fewer panics, fewer moments when people must spend themselves correcting the world by hand. But the warning is there too. When a machine becomes self-correcting, people may trust it too much. Owners may ask more from it. Workers may be told to keep pace with its steadiness. The absence of breakdown may be mistaken for the presence of care. One evening, long after Watt's visit, Samuel found Tom alone near the engine. The governor turned above them, the balls lifting and falling slightly as the load changed in the rooms beyond. "Do you still dislike it?" Samuel asked. Tom listened to the engine before answering.

"No." "Do you like it?" "I like what it prevents." "That is not the same." Tom smiled. "No. But it is enough for many things." They stood together in the heat and noise. Above them, the iron balls spun outward and inward, outward and inward, not tired, not proud, not kind, not cruel. They did not know the men below them. They did not know the women at the looms, the children carrying waste, the owner calculating output, the mothers waiting at home, the injuries avoided, the injuries still to come. They only answered motion with motion. And because they did, the room held its speed. The belts ran. The workers breathed a little easier. The engine, which had once needed a man's hand every time it began to lose itself, now carried within it a small reminder: not so fast.

CHAPTER 4

Control, Navigation, and the Promise of a Visible Coast

They that go down to the sea in ships, that do business in great waters; these see the works of the Lord.

—The Bible, Psalm 107:23–24, King James Version

On a November night in the Baltic, long before satellites could whisper positions from orbit, a Swedish mate named Erik Lund stood on the wet deck of a cargo steamer and tried to turn darkness into geography. The sea had erased the horizon. Rain came sideways. The compass card trembled under its lamp. Somewhere ahead, there should have been rock, channel, shoal, and harbor, but to the eye there was only blackness folded into blackness. The captain, Anders Holm, held the chart under a canvas hood while the helmsman waited for orders he hoped would not be guesses. "There," Erik said at last. "Two flashes, then dark." Anders did not look up immediately. "Are you certain?" "No." "Then look again." They waited. The ship lifted, fell, and shouldered water aside. Then the light returned: flash, flash, pause. Anders touched the chart with a gloved finger. "That is Vinga, or it had better be." The helmsman exhaled. The light

did not pull the ship from danger. It did something more modest and more important. It gave the crew a fact around which judgment could gather.

Navigation at sea is the art of knowing where you are when almost everything around you is moving, hiding, or lying by omission. The water moves. The ship moves. Wind and current move the ship differently from the direction its bow pretends to declare. Fog removes distance. Night removes shape. Snow removes the distinction between air and sea. A chart is a promise written on paper, but the rocks themselves do not glow politely because a sailor needs confirmation. For centuries, lighthouses gave coasts a language. A fixed light said land. A revolving light said this land, not that land. A sequence of flashes became a name. Red and green sectors marked danger and safe water. Buoys, beacons, bells, foghorns, daymarks, and lightships joined the grammar. A sailor did not merely see light. He read it. The coast became legible because engineers learned to turn location into a signal that could survive distance, weather, human fatigue, and fear.

The lighthouse is a control system in a broad and old sense. It does not steer the ship directly, as an autopilot does, but it closes a loop between shore and vessel. The shore sends a known signal. The sailor observes it, compares it with a chart and expectation, corrects the course, and continues. The controller is partly human, partly optical, partly institutional. Someone must decide where a light is needed, what its character should be, how bright it must be, how often it must flash, what color sectors should warn of shoals, who will maintain the lamp, and how mariners will be told what the signal means. A lighthouse that cannot be distinguished from another lighthouse is not merely weak engineering; it is bad language. Imagine Erik, years later, explaining this to a young cadet named Martin. “Why not make every light as bright as possible?” he asks. Erik laughs. “Because a shout is not a sentence.” “Then the flashes are words?” “The flashes are names. At night, names matter.”

Early lighthouses demanded people because flame demanded people. Keepers hauled fuel, trimmed wicks, cleaned soot from lenses, wound clockwork, polished reflectors, watched weather, recorded failures, and lived where comfort was not part of the specification. The great optical improvement came with the Fresnel lens, which gathered and directed light with astonishing efficiency, making lamps visible over much greater distances without simply burning fuel like a desperate bonfire. But even a brilliant lamp remained dependent on attention. A neglected wick smoked. A lens dirtied. A flame went out. A storm delayed supplies. The human keeper was both servant and sensor, actuator and conscience. On an isolated skerry, in an imagined keeper’s house, a girl named Linnea watches her father climb the tower in sleet. “Does the light care if ships come?” she asks her mother. Her mother listens to the wind and says, “No. That is why your father must.”

This dependence on constant human attendance is what Gustaf Dalén helped change. Dalén, the Swedish engineer associated with AGA, developed automatic lighthouse technology using acetylene gas, including inventions that allowed lights to flash efficiently and to conserve fuel. His most famous device, often called the sun valve, used the heat of sunlight to turn the gas supply down or off in daylight and allow the light to operate at night or in darkness. The idea sounds almost simple after it exists, which is the unfairness suffered by many great inventions. The lamp no longer needed a keeper to decide every evening that evening had arrived. Daylight itself became part of the control loop. The environment measured the condition, the mechanism responded, and the beacon spent fuel only when its signal mattered. In 1912, Dalén received the Nobel Prize in Physics for automatic regulators used with gas accumulators for illuminating lighthouses and buoys. He was badly injured in an explosion that same year and lost his sight, a terrible fact made almost unbearably symbolic by his work: a man who helped coasts speak through light would no longer see that light himself.

Let us place Dalén in an imagined workshop conversation shortly before the accident, with an assistant named Karl and an older lighthouse keeper named Oskar visiting from the outer coast. On the bench is a small assembly of metal, blackened rods, valves, and fittings. Oskar looks at it with suspicion. "You are telling me the lamp will know day from night." Dalén smiles. "No. I am telling you the lamp will respond to heat from the sun." "That sounds like knowing if one says it quickly." Karl adjusts a fitting. "It saves gas." Oskar snorts. "Gas is not the only thing out there." "No," Dalén says. "There is also weather, distance, darkness, and the difficulty of sending you supplies in winter." Oskar looks toward the window, but Stockholm offers no answer to an argument about winter rocks and open sea. "And when this clever valve sticks?" "Then it must be designed not to, tested as if it will, and inspected as if it has." Oskar nods slowly. "You speak like a keeper." Dalén answers, "I am trying to make metal behave like one."

The genius of the automatic lighthouse was not that it removed humans from navigation. It moved human labor to another place and another rhythm. Instead of a keeper tending one flame every night, engineers, inspectors, manufacturers, and maintenance crews could build and service many lights that worked unattended for long periods. Remote rocks, buoys, and channel markers could become luminous without requiring families to live in hardship or crews to visit constantly. Flashing conserved acetylene. Automatic shutoff conserved more. Reliability made the system socially useful. A single lighthouse is a local aid; a network of reliable lights is infrastructure. Ships could approach coasts with more confidence. Harbors could extend their working hours. Fishing boats, cargo vessels, ferries, and naval craft could read the night with less dependence on luck. Control entered not as dramatic motion, but as disciplined availability: the right signal, in the right place, at the right time, without a person always standing beside it.

The same principle would matter beyond the shore. In the early age of aviation, pilots did not fly inside the invisible certainty that modern navigation systems create. They followed rivers, railways, towns, coastlines, and, at night, lights. Air routes were built as chains of beacons, fields, signals, and radio aids. Across the United States, night flying over long distances depended on making a continent readable from the air. Beacon lights, including unattended and fuel-efficient designs descended from the same world of automatic illumination, helped turn darkness from a wall into a route. The old phrase "transcontinental night flying" sounds inevitable now, but it was an engineering campaign: place the signals, make them recognizable, keep them burning, maintain them in weather, and teach pilots how to trust them without worshiping them. In an imagined 1920s mail plane over the American plains, a pilot named Ruth Bell scans the dark earth below while her mechanic-passenger, Joe, grips a thermos between his knees. "I lost the last town," he says. Ruth keeps her eyes forward. "Towns sleep. Beacons have better discipline." "You see the next one?" "Not yet." "That is not my favorite answer." Then, far ahead, a point appears, vanishes, appears again. Ruth banks slightly. "There. The country is still where they said it would be."

The connection between lighthouses and night flight is not that Dalén personally invented all aviation beacons, nor that a sea light and an airway beacon are the same object. The connection is deeper and more important for this book. Dalén's work showed how a navigation signal could be made automatic, fuel-conscious, unattended, and reliable enough to join a network. Once that idea exists, it travels. The pilot and the sailor share a problem: movement through a dark or featureless world requires external references that can be trusted. The sea has reefs and headlands; the air has mountains, weather, distance, and fatigue. In both cases, a light is not decoration. It is a controlled promise that the world has been marked for safe passage. Martin, the cadet, might later become a pilot and say to Erik, now old, “In the air, the beacons looked like lighthouses laid across the land.” Erik would answer, “Of course. Humans are very fond of not dying in the dark.”

Modern marine navigation has added layers that would astonish Erik without making his problem obsolete. A ship today may carry GNSS receivers using satellite constellations, radar that paints coastlines and other vessels, AIS transponders broadcasting identity and motion, electronic chart display and information systems that integrate position with digital charts, depth sounders, gyrocompasses, inertial sensors, weather routing, autopilots, vessel traffic services, digital notices, and shore-based monitoring. A bridge can glow with screens that seem to reduce uncertainty to icons. In a modern ferry approaching Rotterdam, Captain Leila van Dijk watches a radar overlay while a junior officer, Tomas, checks the electronic chart. “GPS position agrees with radar,” Tomas says. “AIS target crossing ahead, three cables.” Leila nods. “Visual?” “Not yet.” “Then we do not have him yet.” Tomas looks surprised. “But AIS has him.” “AIS says someone claims to be there, with a transmitter that claims a position. Radar says something reflects energy. My eyes will say whether the night agrees.”

That caution is not nostalgia. It is systems thinking. Satellite navigation can be jammed, spoofed, blocked by equipment failure, degraded by antenna problems, or misunderstood by tired operators. Electronic charts can contain errors or be used at the wrong scale. AIS can be absent, false, delayed, or misread. Radar can be cluttered by rain or sea state. Autopilots can follow a route too obediently if the route is wrong. The more capable the navigation system becomes, the more dangerous it is to treat any one layer as truth itself. Mariners therefore still practice redundancy. They compare satellite position with radar ranges, bearings, depths, visual marks, paper or electronic charts, and common sense. They use traffic separation schemes, pilotage, harbor control, lights, buoys, sound signals, and procedures. The modern bridge is not the end of the lighthouse. It is the lighthouse multiplied into radio waves, databases, screens, alarms, and rules. The old question remains: how does a moving human know where danger is before the hull discovers it physically?

There is also a political and ethical side to navigation infrastructure. A lighthouse is a public good in a particularly literal sense: it shines whether the ship is rich or poor, local or foreign, expected or lost. It does not ask whether the captain has paid enough to see the coast, though governments and harbor authorities may collect dues elsewhere. Modern navigation systems are more entangled with ownership, access, military control, cybersecurity, commercial data, and international standards. Satellite constellations are operated by states or alliances. Digital charts are licensed and updated. AIS data can improve safety but also reveal patterns of trade, fishing, enforcement, evasion, and war. A fishing captain may want rescue services to know where he is and competitors not to know where he found fish. A navy may want signals for itself and confusion for others. A port may want complete visibility of traffic and a smuggler may want darkness returned. Navigation is never only technical. To mark the world is to govern movement through it.

At the end of the chapter, imagine an old lighthouse on a Swedish island now converted partly into a small museum. The lamp room is clean. The lens stands like frozen water. Children climb the stairs and press their hands to the glass. Outside, a modern vessel passes with satellite receivers, radar, AIS, electronic charts, autopilot, and a crew that may glance at the lighthouse only because it is beautiful. Nora, a curator, shows a group the old acetylene equipment and a model of Dalén's sun valve. A boy asks, "Do ships still need lights if they have GPS?" Before Nora can answer, an elderly visitor named Martin, who once served at sea, says, "Need is a dangerous word." The boy turns to him. "But do they?" Martin looks out toward the water. "When everything works, perhaps less than before. When something stops working, you become grateful for every honest light." Nora smiles. "That is why we keep more than one way to know where we are." The sea beyond the window gives no opinion. It has accepted every technology without becoming safe. It accepted fire baskets, lenses, acetylene, radio beacons, radar, satellites, and digital charts with the same indifference. The human answer has been to build layers of correction against that indifference. A lighthouse is one of the oldest and most beautiful of those layers: a controlled signal placed at the edge of danger, saying not come here blindly, not trust the dark, not all is lost. It does not command the ship. It gives the ship a chance to command itself.

Chapter 5

The Grid That Must Balance Every Second

What is the city but the people?

—William Shakespeare, *Coriolanus*, Act III, Scene i

Maya Feld was in an elevator between the twenty-third and twenty-fourth floors of an office building in Manhattan when the afternoon became mechanical in the worst possible way. She was holding a paper cup of coffee she no longer wanted. The elevator stopped as if it had reconsidered its purpose. The lights flickered once, then returned in a weaker form. The fan went quiet. For a moment no one spoke. There were six people inside: Maya, two lawyers still arguing in murmurs, a delivery man with a hand truck, a woman in high-heeled shoes, and a boy of about ten holding his father's briefcase with both hands. "Is this normal?" the boy asked. "No," one of the lawyers said. The other lawyer said, "Elevators stop all the time." "That is a terrible comfort," Maya said. The emergency light gave everyone's face a theatrical hollowness. The delivery man pressed the alarm button. Somewhere beyond the doors, a bell sounded with useless politeness. “My mother is waiting downstairs,” the boy said. His voice did not yet contain fear, only the expectation that adults could arrange reality. Maya envied him for that. “What’s your name?” she asked. “Daniel.” “Daniel, my name is Maya. We are going to be bored before we are in danger.” “How do you know?” She did not know. “Because bored is what elevators do best.”

At almost the same time, in Cleveland, a man named Rick Donnelly stood in a control room and looked at a screen that had stopped telling the truth. Rick is fictional, but his room is not. On August 14, 2003, the electric system in the Midwest and Northeast was being watched by operators who depended on screens, alarms, state estimators, phone calls, procedures, and experience. They were not gods of electricity. They were people trying to understand a machine too large to see. The grid did not sit before them like an engine. It stretched across states and borders, across forests, rivers, suburbs, cities, power plants, substations, steel towers, and wires humming above fields. It was everywhere, which is another way of saying no one could see all of it at once. Rick’s colleague, Denise Alvarez, leaned over her console. “Did you get an alarm for that?” she asked. “For what?” “Line trip.” Rick frowned. “No.” “You should have.” “Then maybe it didn’t trip.” “Or maybe the alarm didn’t.” That distinction was the beginning of trouble. In the hours before the blackout, transmission lines in Ohio sagged into overgrown trees and tripped out of service. Software and alarm problems meant that operators did not fully understand what was happening quickly enough. The system was losing pieces of itself while some of the people responsible for it lacked the picture they needed. In a power grid, ignorance is not empty. It fills with physics. “Call the coordinator,” Denise said. “Already trying.” Rick picked up a phone. Busy. Another line. A voice. Half information. Someone asking whether they were seeing the same thing. No one was seeing the same thing. That was the problem.

Electric power feels simple to the person who flips a switch. It is not simple. At every moment, generation and demand must balance. If people ask for more power, generators must provide it. If a line fails, power flows through other paths, heating them, stressing them, changing voltages. If voltage falls, motors draw current differently. If generators lose synchronism or protective relays sense danger, equipment disconnects to save itself. A grid is not a warehouse of electricity. It is a live negotiation conducted at nearly the speed of light. The negotiation was failing. At 4:10, trains stopped. Computers died. Traffic lights went dark. Air conditioners fell silent in the late-summer heat. In Detroit, Cleveland, New York, Toronto, and hundreds of smaller places, people looked up from desks, registers, hospital beds, subway platforms, kitchens, and factory floors. Some thought the failure belonged only to their building. Then they looked out the window and saw other buildings dark too. In the elevator, Daniel's father pressed the alarm button again. "That makes it ring," the delivery man said. "It does not make them come faster." "Do you work for the building?" "No. I work for sandwiches." The woman in high-heeled shoes laughed once, sharply, grateful for a joke she did not enjoy. Maya took out her phone. No signal. Or too many people trying to use the same invisible channels at once. "I was supposed to be on the subway," she said. "Lucky you," one lawyer said. Maya looked at the elevator doors. "That depends on your definition." Underground, others were less lucky. Subway trains stopped between stations. Passengers waited in heat and darkness while transit workers walked tracks with flashlights. In apartment buildings, people began the long climb up stairwells. In hospitals, backup generators started, and every nurse who heard them understood both relief and warning. A generator is a promise with a fuel tank.

At St. Anne's Hospital, also but assembled from many real hospitals, Nurse Ellen Brooks was checking an infusion pump when the lights dropped. The emergency power came on after a breath that lasted too long. "Generator," said Dr. Patel from the doorway. "I noticed." "ICU?" "Running." "Elevators?" "Ask me something I can answer." A young resident appeared, pale. "Should we move patients?" Ellen looked at him. "Move them where? To the part of the city with electricity?" He swallowed. "Right." "Check oxygen. Check battery backups. Check who is on vents. Then come back and ask better questions." He nodded and ran. In the control rooms, the blackout was becoming a cascade. A cascade is a polite word for a series of protections doing exactly what they were designed to do locally and producing a disaster globally. A line trips to protect itself. Power shifts to another line. That line overloads and trips. Generators disconnect to avoid damage. Frequency and voltage swing. More equipment protects itself. The system separates into islands, some with too much generation, some with too little. What began as a problem in part of Ohio becomes a continental event. Rick watched the map change in ways a map should not change. "We're losing it," Denise said. He wanted to say no. Operators are trained not to surrender to drama. But the screen showed drama without needing language. "How far?" he asked. "Michigan. Ontario. New York reporting trouble." "New York?" "Yes." Rick stared. New York was not a neighboring room. New York was a different world. Yet the grid had made it intimate.

By the time ordinary people understood the scale, scale was already the story. The blackout affected about 50 million people in the northeastern United States and Ontario, taking tens of thousands of megawatts of load out of service. More than 500 generating units at hundreds of plants shut down during the event. The numbers are large enough to become abstract, but abstraction is exactly what electricity had lost that afternoon. Fifty million meant Daniel in the elevator. It meant Ellen checking ventilators. It meant a restaurant owner watching food warm in refrigerators. It meant a man in Toronto directing traffic at an intersection because the lights were dead. It meant a woman on the twenty-eighth floor realizing she had never bought a flashlight because the city had always seemed to provide light as a public habit. At street level in New York, people poured out of buildings. The sidewalks filled with office workers carrying briefcases and shoes, because shoes chosen for elevators are not always shoes chosen for walking home over bridges. Buses were packed. Subways were silent. Strangers exchanged rumors. "Terrorists?" someone asked. "They say Canada." "Canada did it?" "No, it started there." "I heard Ohio." "How can Ohio turn off New York?" That was a good question. The answer was the grid.

Maya and the others were freed from the elevator after almost an hour. A maintenance worker opened the doors from above and told them not to rush. "We have nowhere to rush to," Maya said. Daniel's mother was waiting in the lobby. She held him so tightly he protested, which only made her hold harder. "Mom, I'm fine." "Be fine quietly." Maya stepped outside into a city that had become both calmer and more uncertain. Without traffic lights, drivers negotiated by eye contact and nerve. Without office lights, glass towers reflected a sky that seemed too indifferent. People were walking north in streams. Some were cheerful. Some were afraid. Some were already turning the failure into a story they could tell later. At St. Anne's, Ellen found Dr. Patel in the hall. "Generators are stable," he said. "For now." "You always say for now." "Because I am usually right." "Do you ever trust anything?" She looked toward the ICU. "I trust things that are checked by people who do not trust them too much." The blackout lasted hours for some people and longer for others. In some places power returned the same evening; in others, restoration stretched into the next day or beyond. Ontario asked people to reduce use while generation returned. New York became a walking city. Families slept on rooftops or near windows. People shared candles. Some suffered. Some died in indirect ways: from fire, carbon monoxide, heat, medical vulnerability, falls, and the ordinary hazards that appear when the background system disappears. The official investigation later called the blackout preventable. It identified failures in vegetation management, situational awareness, reliability coordination, operator training, and tools. FirstEnergy's alarm system failure mattered because it left operators without critical warnings. Trees mattered because transmission lines need corridors, not wishes. Communication mattered because no one utility owns the consequences of its own ignorance. Standards mattered because voluntary reliability was not enough for a system whose failures crossed borders.

Weeks after the blackout, Rick and Denise sat in a meeting with investigators. A man from outside the company asked a question that sounded simple because he had practiced it. "When did you know the system was in emergency condition?" Rick looked at Denise. "Which system?" she asked. The investigator paused. "The bulk power system." "Our part of it? The neighboring part? The part the screen showed? The part the alarms should have shown?" "I am asking when you knew." Rick leaned forward. "That is the problem. The grid can be in trouble before anyone has the word trouble on a screen." The room went quiet. Denise added, "And if the people who need to know do not share the same picture, the grid becomes the messenger. By then it is a brutal messenger." That is the control lesson of the 2003 blackout. A power grid is not controlled from one throne. It is coordinated through many control rooms, many companies, many machines, many automatic protections, many human judgments. Local devices protect local equipment. Operators protect their areas. Coordinators try to see the region. The physics sees everything at once. The physics always has the better map. For the public, the blackout was a failure of electricity. For engineers, it was a failure of visibility and coordination. The operators did not need more power in some simple sense. They needed to know which lines were gone, where power was flowing, which voltages were weakening, who else was in trouble, and how close the system was to a cliff. Control depends on measurement, but measurement depends on instruments that work, alarms that sound, models that converge, people who communicate, and organizations that act before the cliff becomes obvious.

Years later, Daniel, the boy from the elevator, might become an electrical engineer. Let us allow him that future. In a university classroom, a professor asks why the grid is hard to control. "Because supply has to equal demand," one student says. "Good," the professor replies. "What else?" "Because power flows where physics sends it, not where contracts say it should." "Better." Daniel raises his hand. "Because people only notice it when it fails." The professor smiles. "That may be the best answer." The amazing thing about the grid is not that it failed on August 14, 2003. The amazing thing is that it usually does not. Every second, generators, loads, transformers, protection systems, markets, operators, forecasts, maintenance crews, tree trimmers, software, and control systems participate in keeping a continent synchronized. Millions of people wake, cook, charge phones, ride elevators, refrigerate medicine, pump water, ventilate patients, trade stocks, heat homes, cool offices, and assume the lights will come on because almost always they do. How often is almost always? Reliability engineers measure this in several ways. One common measure is SAIDI, the average number of minutes or hours of non-momentary interruption, sometimes including planned interruption of service, a customer experiences in a year. In the United States, recent annual averages including major events have often been several hours per customer: nearly five hours in 2019, a little over seven hours in 2021, and about five and a half hours in 2022. Five and a half hours without power in a year sounds like a lot if those hours are yours during a storm, but as availability it is roughly 99.94 percent. If major events are excluded, average outage time can be closer to about two hours a year, which is about 99.98 percent availability. These are rough ways of speaking, and they hide huge differences between places, but they give scale to the miracle.

Maya would not have called it a miracle while trapped in the elevator. Ellen would not have called it a miracle while counting ventilators. Rick would not have called it a miracle while watching a map fail. But afterward, when the lights returned, each of them understood something about the background. Electricity had seemed like a substance delivered through walls. It was really an agreement renewed every instant. The grid is a machine made of machines, but also a promise made by strangers. You trust tree crews you will never meet to cut branches before they touch lines. You trust operators you will never see to notice alarms. You trust protection devices to sacrifice part of the system before the whole is damaged. You trust generators to remain synchronized with other generators hundreds of miles away. You trust engineers who design models, regulators who write standards, crews who repair lines in storms, and neighbors who reduce demand when asked. The promise is not perfect. It never will be. Storms will come. Trees will grow. Software will fail. People will misunderstand. Equipment will age. Heat waves will press air conditioners into the same hour. Ice will pull wires down. Fires will burn corridors. Attackers may try to harm the system. Markets may stress it in ways physics does not forgive. A grid is not safe because it is invulnerable. It is safe because thousands of corrections happen before most of us notice that correction was needed.

On the night of August 14, after walking miles home, Maya climbed the stairs to her apartment with aching feet. The hallway was dark. Someone had placed candles along the landing in jars. A neighbor she barely knew sat on the steps with a battery radio. "Any news?" Maya asked. "Ohio, maybe. Canada, maybe. Trees, maybe. Nobody knows." "That's reassuring." The neighbor laughed. "You want water? I filled pots before the pressure dropped." Maya sat beside him, too tired to refuse. Across the city, people were doing similar things: sharing water, listening to radios, walking instead of riding, discovering which neighbors owned flashlights and which owned only scented candles. The blackout made the grid visible by removing it. It also made other systems visible: kindness, fear, rumor, stairs, heat, patience. When power returned, some people cheered. Others simply slept. Machines restarted with beeps and hums. Refrigerators resumed their low domestic confidence. Elevators came back to life. Traffic lights began ordering intersections again. The city did not become normal all at once. It reassembled itself circuit by circuit, habit by habit. Daniel, safe at home, asked his mother before bed, "Will it happen again?" She wanted to say no. Adults often want to give children false certainty because it feels like kindness. Instead she said, "Probably not tonight." "Why not?" She looked at the lamp beside his bed, newly bright, as if it had returned from a journey. "Because a lot of people are working very hard so it doesn't." That was true enough. The grid deserves that kind of answer. It is not magic. It is not guaranteed. It is not merely wires. It is control as civilization: constant balancing, constant watching, constant correction, mostly hidden, occasionally failing, usually succeeding so completely that success looks like nothing at all.

CHAPTER 6

The Test That Would Not Stop

A little neglect may breed great mischief.
—Benjamin Franklin, *Poor Richard's Almanack*, 1758

In Pripyat, the evening began gently. The children had been outside late because spring had finally made the courtyards usable again. Mothers called from balconies. Bicycles leaned against stairwells. A boy named Misha chased a ball between apartment blocks while his older sister Lena pretended not to watch him and watched him anyway. Their father, Sergei, was on the night shift at the power plant, and their mother, Irina, had left soup on the stove in case he came home hungry before dawn. "He will not be back before breakfast," Lena said. "How do you know?" Misha asked. "Because the plant eats men at night." Misha looked toward the darkening horizon. Beyond the trees and the straight Soviet roads stood the Chernobyl Nuclear Power Plant, lit like a city that did not sleep. Its chimneys and blocks were familiar shapes, almost domestic to the people of Pripyat. The plant gave the town wages, apartments, schools, shops, a swimming pool, a future. Children drew it in school without fear. It was where fathers went in clean work clothes, where buses carried shifts, where electricity was made, where the adult world converted discipline into light. "It does not eat him," Misha said. "Then why is he always tired?" Their mother opened the balcony door. "Both of you inside." "Five minutes!" "Three." "That means five," Misha whispered.

In Unit 4, the night was not gentle. The control room had the look of a place built by people who trusted panels more than faces. Buttons, meters, lights, switches, recorders, alarms, instructions, telephones. The room was full of information and still capable of hiding the thing that mattered. Men sat or stood in the stale brightness of the shift. They knew one another's habits: who smoked too much, who joked when nervous, who became silent, who obeyed quickly, who asked questions only after deciding not to ask them aloud. Leonid Toptunov was young. Too young, some would later say, though youth is often blamed after catastrophe because it cannot defend itself. He was an engineer, trained, serious, and now seated before a reactor that was behaving badly. Alexander Akimov, the shift supervisor, carried responsibility in the shoulders. Anatoly Dyatlov, deputy chief engineer, was present for the test and had the forceful manner of a man accustomed to being obeyed. The test had a good intention. That is one of the hardest facts to hold. They wanted to know whether, if external power failed, the spinning turbine could continue producing enough electricity for a short time to keep coolant pumps running until diesel generators came online. It was a safety test. Not a stunt. Not sabotage. Not madness. A test meant to answer a practical question: in the seconds after a power failure, would the plant have enough momentum to protect itself? Good intentions are not protection against bad conditions. The test had been delayed. The reactor had been held at partial power longer than planned because the electrical grid still needed output. The day shift had prepared one situation; the night shift inherited another. Xenon had built up in the reactor, poisoning the reaction and making power difficult to manage. The planned operating state had slipped away hour by hour, but the test remained, waiting like an obligation no one wanted to abandon.

Toptunov watched the numbers. "Power is falling," he said. Akimov stepped closer. "How far?" Toptunov answered, and the number was too low. For a moment the room seemed to become smaller. Procedures were written for a world in which people stopped when limits were crossed. Real rooms contain fatigue, hierarchy, embarrassment, schedule pressure, pride, and the fear of telling a superior that the night's work must be abandoned. "We should shut it down," someone said. Later, people would argue who said what, who insisted, who resisted, who understood. Disaster turns conversations into evidence. But in the room, at the time, they were only men trying to decide whether a failed preparation could still be rescued. Dyatlov's voice cut through the uncertainty. "Raise the power." Toptunov hesitated. "Comrade Dyatlov, it has fallen too far." "Raise it." Akimov looked from the panels to Dyatlov. A shift supervisor lives between instruments and authority. The instruments have numbers. Authority has a voice. "Leonid," Akimov said quietly, "withdraw rods." Control rods are a beautiful phrase for a frightening responsibility. They enter the reactor to absorb neutrons and slow the chain reaction. They withdraw to allow the reaction to grow. In a well-behaved reactor, they are the hands by which humans ask the atom to remain useful. That night, too many were withdrawn. The reactor was being pulled into a state the rules were meant to forbid. Toptunov's hand moved. The plant obeyed, in its own dangerous way. "Power is coming up," he said. "How much?" "Not enough." "Continue." There were alarms. There were signals. There were procedures that should have had the authority of law. But rules do not enforce themselves. They live only in people, and people can be pressured, tired, mistaken, obedient, afraid of failure, or convinced that one more exception will be harmless because many previous exceptions have been harmless.

In Pripyat, Irina put Misha to bed. "Will Papa bring bread?" he asked. "If the shop has bread." "Will he wake me?" "No." "Tell him to." "Sleep." Lena stood in the doorway. "Mama, why do they test things at night?" Irina tucked the blanket around Misha. "Because electricity does not sleep." "That is not an answer." "It is the answer I have." At the plant, the test continued to assemble itself from compromises. Some safety systems had been disabled or bypassed for the purpose of the test. The emergency core cooling system had been made unavailable earlier. Automatic shutdown signals were blocked so the test would not stop prematurely. Pumps were running in configurations that affected coolant flow. The reactor, at low power, with many rods withdrawn, was not in the calm state a reactor should be in before humans ask it experimental questions. Akimov rubbed his face. "We are outside the instructions," Toptunov said. Dyatlov turned sharply. "The instructions do not run the plant. We do." That sentence, whether spoken exactly or only true in spirit, is the temptation at the heart of unsafe systems. The operators were not fools. They were not villains standing before a machine they wished to destroy. They were trained men inside an organization where production, authority, secrecy, and overconfidence had worn grooves into behavior. They had learned, as people in many dangerous industries learn, that rules are absolute until a superior treats them as negotiable. "If we stop now," Akimov said, "there will be questions." Dyatlov looked at him. "There will be questions if we do not complete the test." "There will be questions if we complete it in the wrong state." For a few seconds no one answered. In those seconds lay an alternate history. Then the work resumed.

The test began at 1:23 in the morning. Steam to the turbine was shut off. The turbine began to coast down, spinning from momentum, just as the experiment required. The pumps changed their behavior as the electrical supply from the slowing turbine changed. Flow changed. Steam voids formed in the reactor channels. In an RBMK reactor, this mattered terribly. More steam could mean more reactivity. More reactivity meant more heat. More heat meant more steam. A loop had been waiting inside the design. Toptunov saw the power begin to rise. "Power is increasing." Akimov leaned in. "How fast?" The answer came in numbers, but the meaning was simple: too fast. "Shut it down," Akimov said. The AZ-5 button was pressed. In a different reactor, in a different condition, this would have been the final human command for safety: insert the rods, stop the reaction, end the test. Men had gone too far, and now the machine would be told to halt. But Unit 4 was not in a forgiving condition, and the control rods themselves carried a hidden cruelty. Their graphite tips initially displaced water in parts of the core, and under the conditions of that night this could increase reactivity before reducing it. The shutdown button, the symbol of final protection, became part of the surge. Toptunov had time to understand almost nothing. The reactor power rose with impossible speed. "What is happening?" someone shouted. Akimov stared at indications that no longer belonged to ordinary operation. There was a sound from below or within, a deep violence moving through concrete and metal. Then another.

The reactor was destroyed. The roof was blown open. Graphite and fuel were exposed to the night. Fire began where no ordinary fire should be. The plant had become not only damaged but open, its invisible contents entering the world.

In Pripyat, windows flashed. Misha woke. "Thunder?" Lena was already at the window. "Mama." Irina came from the kitchen and saw the glow beyond the trees. At first, the mind tries to protect itself by choosing a familiar explanation. A transformer. A fire. An industrial accident of the ordinary kind. Not the reactor. Not the thing no one is supposed to imagine. "Get away from the window," she said. "Is it Papa's block?" Lena asked. Irina did not answer quickly enough. At the plant, men ran toward disaster because that was their work. Firefighters arrived without being told the truth of what burned. Vasily Ignatenko, one of the real firefighters whose name would later become inseparable from the human cost of Chernobyl, went with others into a scene that looked like fire and was also radiation. They climbed, carried hoses, stepped near graphite fragments that were not stones, breathed a danger that had no smell, no heat proportional to its harm, no honest warning to the senses. "Careful there!" one man shouted. "With what?" another answered. No one had the right word yet. Inside, Akimov and Toptunov did not flee into innocence. They tried to feed water to a reactor that was no longer intact. They remained in a story that had already judged them and still required their bodies. Later, blame would gather around names. Some blame was deserved. Rules had been violated. A reactor had been taken into a forbidden state. Warnings had been ignored or overridden. But blame alone is too small for Chernobyl. Akimov reportedly insisted, "We did everything right." It is one of the most haunting sentences in technological history, because it can be read in two opposite ways. It can mean denial. It can mean confusion. It can mean that a man who knew rules had been bent still could not reconcile his actions with the scale of the result. No one presses a button expecting to open the core to the sky.

By morning, people in Pripyat came outside to look. They stood on balconies. They walked with children. Some watched the glow. Some spoke of an accident, but the word accident was still too small. Radiation moved silently through clothes, hair, lungs, grass, dust, milk, shoes. Children went to school.

Weddings were photographed. Buses had not yet come to take the city away. Misha wanted to go outside. "No," Irina said. "Everyone is outside." "Not you." "Why?" Because your father has not called. Because the plant is burning. Because I do not know what the air contains. Because no one tells us the truth in time. She said, "Because I am your mother." That answer held for ten minutes. The evacuation began the next day. Buses arrived. People were told it would be temporary. Take documents. Take necessary things. Leave pets. Close windows. The instructions sounded administrative, almost calm, as if the state could make catastrophe orderly by speaking in lists. Lena packed a book, two photographs, and a sweater. Misha tried to bring a toy truck. "No," Irina said. "It is small." "No." "Papa gave it to me." She took the truck from his hand, then gave it back. "Put it in your pocket." "Will Papa come on the bus?" Irina looked toward the plant. "He will come later." The lie was not chosen. It rose by itself because the truth had no shape she could give a child. Chernobyl is often told as a story of operators who broke rules, and it is that. They had a good intention: to test whether the turbine's remaining energy could help keep coolant flowing until backup generators came online. But they continued after the reactor had fallen into a dangerous low-power state. They withdrew too many control rods. They disabled or blocked safety systems. They carried out an experiment in conditions that should have stopped the experiment. Human beings pushed the plant past its rules, and the plant answered catastrophically.

But Chernobyl is also the story of rules that were not strong enough against culture. A rule written in a binder is weak if a superior can make a young operator feel that stopping is cowardice. A safety system is weak if it can be disabled casually for a test. A control room is weak if it does not make the true danger visible. A design is weak if, under certain conditions, the emergency shutdown can initially worsen the reaction it is meant to stop. A culture is weak if success means completing the task and safety means hoping the task forgives you. This is where control engineering becomes more than automatic regulation. Control is not only rods, pumps, valves, sensors, and buttons. Control is also procedure, authority, information, training, permission to stop, and the design of a room in which one frightened junior engineer can say no and be heard. Imagine, years later, an engineer named Natalia speaking to students who have come to study safety systems. She is fictional, but the classroom is real enough. On the screen behind her is a diagram of an RBMK reactor. The students expect physics. She gives them a conversation. "What caused Chernobyl?" she asks. A student raises his hand. "Operator error." "Good. What else?" "Design flaws." "Good. What else?" "The positive void coefficient." "Good. What else?" Another student says, "The control rods." "Good. What else?" The room grows irritated. Students like causes to end. Finally a young woman says, "Pressure?" Natalia nods. "Pressure. Delay. Pride. Bad information. A weak safety culture. A test that should have been stopped. A machine that did not forgive being used outside its rules. If you name only one cause, you make the disaster smaller than it was." "But the operators broke the rules," the first student says. "Yes," Natalia says. "Never forget that. But never stop there. If your explanation ends with bad operators, your next plant will wait for better people. Engineering cannot be built on the hope that people will always be better than the system around them."

That is the lesson Chernobyl gives this book. The clock taught us that a useful system must discipline its own behavior before it can discipline the world around it. The thermostat taught us that a room should not have to wait for discomfort to become complaint. The governor taught us that power needs restraint before speed becomes danger. The grid taught us that large systems fail when visibility and coordination break down. Chernobyl teaches something darker: a safety-critical process must be controlled not only when everyone obeys, but especially when tired, pressured, hierarchical humans begin to make exceptions. In the control room, the operators believed they were conducting a test. In the reactor, physics did not care what they believed. That sentence is cruel, but safety engineering begins by accepting it. Nature does not grant moral credit for good intentions. A reactor does not know that a test is meant to improve safety. A pump does not know that a manager wants completion. A neutron does not know that a young engineer is afraid of contradicting a superior. The system responds to conditions, not motives. After Chernobyl, the world counted cancers, evacuations, contaminated land, dead firefighters, sick workers, abandoned homes, political consequences, design changes, and the long shadow of mistrust. Pripyat became a city of absence. The amusement park never opened properly. Apartments held the shapes of interrupted lives. Toys remained where children had left them. The plant that had promised a future gave the town a museum of stopped time. Years later, Misha, now grown, might return as part of a controlled visit. Let us allow him that. He stands below the apartment where his mother once called from the balcony. Trees have taken space from the pavement. Paint has peeled. The air is measured before he enters, because even memory now requires instruments. "Was this your building?" the guide asks. Misha nods. "Do you want more time?" He looks up at the balcony. "No. I want less."

In his pocket, impossibly, he carries the toy truck. Not the original perhaps; memory is sometimes kinder when it lies. But in this story, let it be the original. His father gave it to him. His mother let him take it. A child carried it onto a bus because adults said the evacuation would be temporary. Control failed at Chernobyl in rods and pumps and procedures. It also failed in speech. The town was not told quickly enough what had happened. Firefighters were not fully protected from what they faced. Operators were not given a design that made danger clear enough or forgiving enough. Rules existed, but they did not hold against the organization that bent them. Misha turns the truck in his hand. "They said we would come back," he says. The guide says nothing. There are disasters that end when the fire is out. Chernobyl was not one of them. Its control loop extended into decades: monitoring, exclusion zones, sarcophagi, medical studies, political collapse, engineering reforms, arguments about numbers, arguments about blame, arguments about memory. The reactor was destroyed in seconds. The consequences kept operating. So if this chapter has a human voice, it is not only the shout in the control room or the firefighter on the roof. It is also the mother packing a bag, the child asking about his father, the engineer years later refusing a simple answer, the adult returning to a town that became a warning. The experiment had a good intention. That is why it frightens me. Bad intentions are easy to condemn from a distance. Good intentions invite us closer. They ask whether we, too, might continue a test because it was important, because stopping would be embarrassing, because the schedule mattered, because others had accepted the risk, because the rule seemed conservative, because the machine had tolerated exceptions before. Chernobyl says: do not confuse previous forgiveness with safety. The plant had operated. The men had experience. The system had survived other compromises. That history may have made the final compromise feel less like a cliff. But the edge was there whether they saw it or not.

Control engineering, at its best, builds systems that make the edge visible, systems that resist being pushed toward it, systems that give people permission to stop before heroism is required. It does not assume perfect obedience. It assumes fatigue. It assumes pride. It assumes fear. It assumes that someone, someday, will try to do the useful thing in the wrong condition. And then it asks: will the system save them from that mistake, or will it amplify it? At Chernobyl, the system amplified it.

Chapter 7

Control, Hidden in the Sky

The natural function of the wing is to soar upwards and carry that which is heavy up to the place where dwells the race of gods.

—Plato, *Phaedrus*, 246d, trans. Benjamin Jowett

Stockholm had given itself to summer, and Elin had decided that her grandfather was walking too slowly. "Morfar," she said, pulling at his hand, "if we don't hurry, we won't see anything." Gunnar Larsson allowed himself to be pulled. He was seventy-one, retired from the Air Force, and careful with his knees on cobblestones. He had once known the sound of aircraft by type before he could see them. Now his granddaughter knew them mostly as shapes from posters and television. She was eight years old and had no respect for old joints. "The sky is large," he said. "It is difficult to miss." "But people stand in front." "People rarely stand in front of the sky." Elin looked back at him with the impatience of a child surrounded by adults who insisted on being literal. Around them, the Stockholm Water Festival moved in bright disorder. It was August 1993. The city smelled of water, warm stone, frying batter, coffee, sunscreen, and exhaust from boats. Music came from somewhere across the crowd, cheerful and badly mixed. A man was selling balloons. A woman laughed with her head tilted

back. Teenagers leaned against a railing pretending not to care about anything. Tourists photographed the water as if Stockholm had arranged it especially for them. At a kiosk near the bridge, Karin bought two paper cups of lemonade and tried to balance them with a bag of cinnamon rolls. Her husband, Olof, held their baby against his shoulder. The baby was asleep, mouth open, one cheek pressed flat against Olof's shirt. "You carry her," Karin said. "I'll carry civilization." "Is that what cinnamon rolls are?" "In Sweden, yes." Beside them, two women were deciding where to stand. Marion wanted a better view. Barbro wanted shade. "You always choose shade," Marion said. "Because I plan to survive the afternoon." "We came to see an aircraft." "Aircraft are in the sky. Shade is on the ground. We can have both."

Marion laughed and let herself be persuaded. They found a place where they could see the open sky between trees and rooftops. It was close enough to feel part of the crowd, but not so packed that they could not move. Later, both would remember the place differently. Memory after fear does not keep maps clean. Gunnar and Elin came to a stop not far from them. Elin immediately climbed onto a low stone edge. "Down," Gunnar said. "I can't see." "If you fall, you will see the hospital." She stepped down, then looked up at him. "Did you fly one like this?" "No. This one is new." "But you flew fighters?" "A long time ago." "Were they as fast?" Gunnar smiled. "Every generation thinks its aircraft are fast." "That means no." "That means eat your roll." She took the cinnamon roll he had bought her and held it with both hands, eyes already searching the sky. Gunnar followed her gaze. He felt the old mixture in his chest: pride, suspicion, grief. A fighter display over a city was meant to produce pride. It did produce pride. But Gunnar had buried too many friends to look at any military aircraft with only admiration. Olof shifted the sleeping baby. "Is it safe to fly so low here?" Karin asked. "They would not do it if it wasn't safe," he said. "That is a very trusting sentence." "Today I am a trusting man." "Since when?" "Since carrying the baby. It makes arguing harder."

The sound arrived before the aircraft. Conversations thinned. Heads turned. Elin grabbed Gunnar's sleeve so hard he felt her fingers through the cloth. The JAS 39 Gripen came into view, sharp and gray, a new Swedish shape against the summer sky. People pointed. Someone cheered too early. The aircraft passed with the astonishing indifference of speed: here, then elsewhere, leaving sound to catch up behind it. "There!" Elin shouted, though everyone had seen it. "I see," Gunnar said. "Is that the Gripen?" "Yes." "It looks small." "It is not small." "It looks like a paper plane." Gunnar watched the aircraft turn. "Paper planes are stable because children forgive them." Elin frowned. "What does that mean?" "Nothing. Watch." But it did mean something. Gunnar knew enough to know that the new aircraft was not like the machines of his youth. The Gripen was agile because it was unstable. It used computers to remain flyable, computers between the pilot's hand and the aircraft's surfaces, computers making small decisions faster than a body could feel them. Fly-by-wire, they called it, as if the phrase were simple. A pilot gave commands. The system interpreted them. The aircraft answered. When everything worked, the result was not hesitation but grace. The crowd saw a clean movement in the sky; Gunnar knew there was an invisible conversation inside it.

No one in the crowd saw the computers. They saw a fighter. Marion shaded her eyes. "Beautiful," she said. Barbro looked at the same aircraft and said, "Too close." "You think everything is too close." "That is why I am still alive." The Gripen maneuvered above the city. For another moment, it belonged to the festival: a planned astonishment, a national machine, a controlled danger offered as entertainment. Lars Rådeström, the pilot, was invisible inside the cockpit, but the crowd lent him trust. They had to. Every spectator at an air display enters an agreement with a person they cannot see: I will stand here, and you will keep that machine in the sky. Then the agreement broke. At first, only a few understood. Gunnar did. His body knew before his thoughts did. The aircraft's attitude changed in a way that did not belong to display. The nose rose. The motion lost its clean line. The correction seemed to call for another correction. The Gripen no longer looked like a machine choosing a path. It looked like a machine arguing with itself: pilot, computer, control surfaces, and air answering one another too late or too much. "Morfar?" Elin said. Gunnar put his hand on her shoulder. "Down." "What?" "Down from the stone. Now." She obeyed because his voice had changed. Karin heard the change in the crowd before she saw the aircraft. There is a sound people make when admiration turns into fear. It is not a scream at first. It is a collective intake, a question made of breath. "Olof," she said. He turned with the baby still asleep against him. "What?" "Move." "Where?" That was the terrible question. Where does one move when the danger is falling from above? The aircraft was not a car with a lane, not a horse with a direction, not a boat on water. It was speed and mass descending through air, and the crowd had only guesses. Marion saw it coming wrong. "Barbro." "I see it." "No, Barbro, move."

People began to run. Some ran toward the water. Some ran away from it. Some ducked. Some froze because the mind cannot always convert danger into instruction quickly enough. A man dropped an ice cream and did not notice. A child cried because adults were suddenly frightening. Someone shouted, "Spring!" Run. Someone else shouted, "Ner!" Down. Gunnar pulled Elin close and crouched beside a low wall. He did not know if it was the right choice. He only knew he had to make one before certainty arrived, because certainty would arrive as impact. Above them, the pilot ejected. The seat fired. The canopy separated. A human being left the aircraft and became, for the crowd, a second falling object, then a parachute. Elin saw the white of it. "The pilot!" she cried. "Don't look at him," Gunnar said. "Look at me." But she looked anyway, because children look at the thing adults tell them not to look at. The parachute opened. The man was alive. But the aircraft was still coming down.

The Gripen struck Långholmen near the Västerbron area with a violence that seemed too large for the sunny day. Fire opened. Smoke rose. The sound hit bodies and stone. Heat came after it. The crowd broke fully now, not as spectators but as people trying to account for their own limbs, their children, their friends, their bags, their elderly parents, their route away from flame. Karin stumbled and recovered. Olof turned his shoulder around the baby, making of his own body a foolish and necessary shield. Marion fell to one knee. Barbro pulled her up with both hands. "Up, up, up!" Barbro shouted. "I'm up!" Marion shouted back, though she was not. A woman nearby screamed and slapped at her hair. Heat had touched her. For a moment people moved away from her because fire makes a private space around itself. Then someone moved toward her. Karin saw it happen: a young man with festival sunglasses, absurdly bright blue, grabbed for water and found none. Olof thrust the baby's bottle toward him without speaking. The young man stared at it. "Use it!" Karin shouted. He did. The bottle, meant for milk or water or the small domestic emergencies of an infant, became part of an aircraft accident. The detail would stay with Karin longer than the smoke: her baby's bottle in a stranger's hand, water poured against heat from a fighter jet. "Is she all right?" Olof asked. "I don't know." "Is the baby?" Karin looked at the child, who had finally woken and begun to cry. "Now she is."

Sirens came. Police voices cut through festival noise. People were pushed back. Others resisted because someone they loved was on the other side of the pushing. The aircraft burned. The parachute hung where people could see it. Rumor spread in several directions at once. "The pilot is dead." "No, he ejected." "It hit people." "How many?" "I don't know." "It almost hit us." Almost became the word of the afternoon. Almost hit us. Almost killed her. Almost went into the crowd. Almost a catastrophe beyond counting. Later, the official facts would be clearer. Lars Rådeström survived. No one on the ground was killed. One woman was injured by burns. The aircraft was destroyed. Investigators would examine the accident, the maneuver, the flight control system, the relation between pilot input and aircraft response. The Gripen program would continue. The aircraft would be modified, matured, trusted again. But official facts arrive too late to govern the body. That evening, Marion sat at a kitchen table and held a cup of coffee she did not drink. "How close?" someone asked her. She looked at the cup. "Close enough." "But how many meters?" She shook her head. "Meters are not the right measure." Barbro, at home, found that her clothes still smelled of smoke. She put them in the wash and then took them out again because some irrational part of her wanted proof. Her husband asked whether she was hurt. "No." "You are shaking." "That is not the same thing."

Gunnar walked Elin home in silence. She had stopped asking questions, which worried him more than questions would have. At the door she said, "You knew before everyone." "No." "You did. You told me to get down." He hung his jacket slowly. "I knew something was wrong." "Because you were a pilot?" "Because I have seen aircraft stop behaving like aircraft." She considered this. "Can machines decide to stop?" "No." "Then why did it?" Gunnar sat down before answering. "It did not decide. Something in the conversation failed." "What conversation?" "Between the pilot and the aircraft. Between the aircraft and its computers. Between what was asked and what happened." Elin looked unconvinced. "Machines talk?" "All the time. Not with words." "Then how do we know what they say?" Gunnar looked toward the window, where the evening light had softened the city. "Sometimes we know because they answer correctly. Sometimes we know because they don't."

The next morning, in a conference room far from the smell of smoke, engineers began the slower work of making the accident speak. Let us place two engineers there, because reports can make it seem as if machines investigate themselves. They do not. People do. Eva Lund had worked on control systems long enough to distrust both panic and reassurance. She arrived with a notebook, a pencil, and the particular exhaustion of someone who knows that public fear is already moving faster than technical understanding. Across from her sat Nils Ek, younger, pale from too little sleep. On the table were early printouts, timelines, witness notes, and coffee in paper cups. They did not yet have the answer. That matters. Accidents often look obvious only after the vocabulary has arrived. "The newspapers want a cause," Nils said. "Of course they do." "Do we have one?" Eva looked at the papers. "We have wreckage. We have recordings. We have memories. We have a sequence we do not yet understand." "That is not what they want." "It is what accidents usually give us first." He rubbed his eyes. "Pilot error? Software? Control law? Display maneuver?" "Do not make a menu out of it," Eva said. "If we choose too early, we stop listening."

Weeks later, after simulations had been run, data compared, timelines rebuilt, and arguments repeated until they became less dramatic and more useful, the shape of the accident became clearer. Nils found Eva in the same conference room, now with more folders and less coffee. "Pilot-induced oscillation," he said, not as a guess this time but as a phrase that had become heavy from use. Eva nodded slowly. "Yes. But be careful with it." "Because people will hear pilot mistake." "Some will. Others will hear computer madness. Neither is enough." "Then what do we say?" "That the pilot and aircraft formed a loop, and the loop began to amplify instead of settle." She tapped the paper. "The pilot commands, the system interprets, the aircraft responds, the pilot senses, and the loop continues. Usually so smoothly that everyone calls it flying. In a pilot-induced oscillation, that loop loses its rhythm. The pilot is not outside the system correcting it like a person straightening a picture on a wall. He is inside the system. His correction changes what he sees next. If the aircraft answers with delay, or with more response than the pilot expects, the next correction can arrive at exactly the wrong moment." "Like pushing a swing?" Nils said. "Yes. Except the swing is a fighter aircraft over a city." "And when the rhythm goes wrong?" "Then everyone discovers the loop was there."

This was the hidden technology. Not the wing, though the wing mattered. Not the engine, though the engine mattered. Not the pilot, though the pilot mattered intensely. The hidden technology was the relation: sensor, software, surface, body, expectation, delay, limit, correction. Pilot-induced oscillation is a failure of that relation. It does not mean the pilot decides to shake the aircraft. It means the pilot's attempt to damp the motion becomes part of the motion. The Gripen was an unstable aircraft made usable by control. Its agility was not a rejection of control but a dependence on it. Without the hidden system, the aircraft could not safely offer the performance that made it impressive. But if the human and the control system misunderstood each other for even a few seconds, agility could turn into amplification. Nils stared at a plot. "Do you think people will understand that unstable can be safer?" Eva almost smiled. "No." "Should we say it anyway?" "Carefully." "It sounds absurd." "Many true things do when compressed." "An aircraft that needs computers to fly can be safer than one that doesn't?" "If the computers are good, if the design is good, if the training is good, if the testing is honest, if the organization learns, if maintenance holds, if pilots understand the machine, if the machine gives them the right cues. Safety is not a property. It is an arrangement." Nils wrote that down. "Do not quote me," Eva said. "Too late." She let him have the joke.

In the days after the crash, Stockholm told the story in pieces. A woman nearly hit. A woman burned. A baby bottle used to cool injury. A pilot descending by parachute. A fireball near a summer crowd. These stories traveled because they had voices. They had faces. They had the force of almost. Statistics did not travel as easily. Gunnar knew the statistics, at least in the rough way old pilots know the dead. When the newspapers called the Gripen dangerous, he did not dismiss the fear. He had seen the fire. Elin had seen it. Fear had earned its place. But he also remembered another Sweden, the Sweden of the late 1950s and early 1960s, when military flying killed so often that grief seemed built into the profession. A week after the crash, he met an old squadron friend, Lennart, for coffee. Lennart had a cane, a white beard, and a laugh that arrived without warning. "They say modern aircraft are too complicated," Lennart said. "They are complicated." "They say the computers are dangerous." "They can be." "You are very agreeable today." Gunnar stirred his coffee. "A fighter fell into a festival. It is a poor week for sounding clever." Lennart nodded. Outside, buses moved through ordinary traffic. "Do you remember F 8 in those years?" "I remember too much." "Funerals every season, it felt like." "Sometimes more." "And those aircraft were not full of computers." "No." "Very pure. Very direct.

Very capable of killing you honestly." Gunnar looked at him. "That is a bitter sentence." "It is an old man's privilege." They sat quietly. Both knew men who had died in aircraft that required no software to be lethal.

The numbers were stark. One summary of the Swedish Air Force during the Cold War gives 625 deaths in flying accidents between 1946 and 1991. The averages are even more startling: about 21 deaths per year in the 1950s, about 13 per year in the 1960s, and still about 7 per year in the 1970s. Of the 661 Saab 29 Tunnan aircraft built, roughly one third were lost in crashes, and 99 pilots were killed in military practice flights in Sweden. The Lansen, too, became associated with catastrophe; in 1960 alone, one account describes 27 military deaths connected with Lansen accidents. "Try telling that to someone who ran from the Gripen," Lennart said. "It would sound like an excuse." "But it is not." "No. It is perspective." "People don't want perspective when they are afraid." "Nor should they be required to." That was the difficult truth. Marion's story and Barbro's story were true in a way no statistic could correct. The aircraft nearly struck them. A woman was burned. A festival became a disaster site. If someone said, "But aviation is safer now," too soon, the sentence would be cruel. A graph does not stand between a person and heat. A trend line does not tell a child where to run.

But the statistics were also true. Without them, the public would misunderstand danger. The old aircraft, more direct and less computerized, had not belonged to a golden age of safety. They had belonged to an age of terrible losses. Modern control systems, simulators, redundancy, flight envelopes, software limits, maintenance procedures, and accident investigations had not removed risk, but they had changed its scale. The fact that the Gripen crash shocked the country was itself evidence of a changed world. In an earlier era, crashes were more common, even if each remained unbearable to those touched by it. At the investigation table, Nils found himself saying something similar. "The accident has a voice," he said. Eva looked up. "What?" "The accident. The witnesses. The fire. The woman with burns. The pilot under the parachute. All of that speaks loudly. But the flights that land safely do not speak." "No." "The corrections that work do not speak." "No." "The accidents prevented by control laws do not speak." Eva closed the folder in front of her. "That is why we count."

"Because statistics speaks for what did not happen?" "Statistics tries to. Poorly, but it tries." He wrote another sentence in his notebook: Statistics does not have a voice. Disasters have voices. Eva read it upside down. "Keep that," she said. A humane engineering culture needs both. It needs Marion saying, "It was close enough." It needs Barbro remembering smoke in her clothes. It needs Gunnar's hand on Elin's shoulder when the aircraft's motion became wrong. It needs Karin shouting, "Use it!" as a baby's bottle became an emergency tool. These stories prevent engineers from hiding inside averages. They remind everyone that a single failure can fill an entire human life.

But it also needs the averages. It needs the cold columns of numbers showing that some eras killed pilots with horrifying regularity. It needs to know that one spectacular modern crash does not by itself prove that hidden control has made aircraft more dangerous. It needs to count the flights that did not crash, the pilots who returned, the crowds that went home with only sunburn and photographs, the invisible successes of systems no one applauded because nothing happened. Elin did not learn all this at once. Children rarely receive perspective on schedule. For weeks she drew aircraft with parachutes. Sometimes she drew fire. Once she drew a plane with a speech bubble. "What does it say?" Gunnar asked. She had written the words carefully: I am trying. He sat beside her. "Is the plane saying that?" "Or the pilot. Or the computer." "Which one?" She shrugged. "All of them." Gunnar looked at the drawing longer than she expected. "That may be right." The Gripen over Stockholm was not one thing. It was pilot, aircraft, software, organization, history, ambition, and public trust joined in a single visible arc. The accident was not best understood as a lone hand making a bad movement, nor as a machine acting alone. It was a pilot-aircraft oscillation, a coupled system losing calm. When it worked, all of that became a graceful machine in the sky. When it failed, the hidden joins showed. That is what control often is: the hidden joining of things that would otherwise not safely belong together. Human intention and unstable aircraft. Power and restraint. Speed and public space. Innovation and trust. Statistics and stories.

On the day of the festival, the story had the louder voice. It still does. A summer crowd. A child looking up. A woman pulling her friend. A pilot under a parachute. Fire near the water. A baby bottle in a stranger's hand. The statistics sit behind that story, quieter but necessary. They say that military flying once killed far more often. They say that modern control, for all its opacity and failure modes, is also part of the long reduction of death. They say that the hidden technology is not innocent, but neither is it the villain. It is one of the ways danger has been made rarer. The final lesson is not comfort. It is tension. If we listen only to the disaster, we may fear the wrong future. If we listen only to the statistics, we may forget the people under the falling machine. Control engineering lives between them. It must make the numbers better without ever forgetting that, when control fails, the result does not fall onto a spreadsheet. It falls into someone's summer afternoon.

Chapter 8

The Rocket That Balances on Fire

To confine our attention to terrestrial matters would be to limit the human spirit.

—Stephen Hawking, *A Brief History of Time*, Chapter 8

The first time Karl Weiss saw a rocket rise from Peenemünde, he forgot to breathe. It was not beautiful at first. Beauty came later, after distance and propaganda and memory had cleaned the image. Up close it was noise, frost, fuel, shouted orders, cables, wind off the Baltic, men with clipboards, soldiers with authority, and a white body standing upright as if it had no business belonging to Earth. Karl was fictional, but the young engineers around him were real enough: men trained in mathematics, mechanics, chemistry, radio, gyroscopes, valves, combustion, and ambition. They had come to the Baltic coast because Germany was at war, and because the army wanted weapons that could travel farther than artillery and arrive faster than warning. Wernher von Braun stood among them, not yet the American television figure with models of moon rockets, but a young German engineer in a military world. Walter

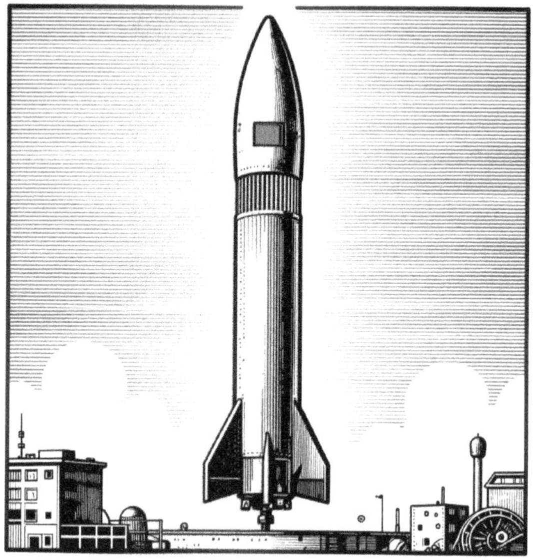

Dornberger, the army officer who commanded the program, moved through the site with the impatient energy of a man who needed results, not dreams. The rocket was called the A-4 before the propaganda name V-2 made it a weapon of vengeance. To some of the engineers it was also, secretly or not so secretly, a step toward space. That doubleness is part of the moral difficulty. The same equations that point upward can be aimed at cities. "Will it fly today?" Karl asked. His friend Ernst kept his eyes on the gantry. "It will leave the ground." "That is not the same thing." "No. But it is the first requirement."

A rocket is a strange promise. A bridge promises to stand still. A train promises to stay on rails. A ship promises to float. A rocket promises to fall upward without losing its mind. It has no road, no track, no wing that can carry it gently if the engine pauses. It balances on violence. It must point in the right direction while burning fuel at a furious rate, while its mass changes, while vibrations pass through its body, while air thickens and thins around it, while a small error in angle tries to become a large error in destiny. "What keeps it straight?" a soldier asked behind them. Ernst smiled without turning. "Faith." Karl said, "Gyroscopes." The soldier frowned. "Which answer is official?" "Gyroscopes," Karl said. "Faith is for after ignition." Inside the rocket, instruments tried to do what no human hand could do from the ground quickly enough. Gyroscopes sensed attitude. Control systems moved vanes in the exhaust and fins in the air. The rocket had to correct itself because by the time a man saw the error, the error would already be history. Guidance was not decoration. It was the difference between a rocket and an explosion with direction. The countdown reached the point where conversation ended. Flame appeared below the rocket, then thick smoke, then a force that seemed to push sound into every rib. The vehicle trembled. For one awful moment it seemed pinned to Earth by its own fury. Then it rose. Slowly first, too slowly for Karl's heart, then faster. Men looked up with faces lit by fire and terror. The rocket leaned slightly. The correction came. It straightened. It climbed. "There," Ernst whispered. "It is listening." Karl did not answer. The rocket was no longer a machine beside them. It was an argument with gravity disappearing into cloud. A few minutes later, someone began to cheer. Someone else did not. Dornberger wanted performance. Von Braun wanted performance too, but also altitude, range, possibility. The army wanted a weapon. The engineers wanted a working machine. Germany wanted victory, or so the men in uniform said. None of these wants canceled the others. They combined into the rocket's flame.

That is where any honest chapter about von Braun must begin: not with Apollo, not with the Moon, not with schoolchildren watching launches on television, but with war. Peenemünde was not a neutral temple of science. The V-2 was built for Nazi Germany. It was fired at cities. Its later production used forced labor under horrific conditions at the underground Mittelwerk and the Mittelbau-Dora camp. More people died building the weapon than were killed by its explosions. No sentence about space exploration should be allowed to pass over that fact too quickly.

One evening, Karl and Ernst stood outside a barracks after a test failure. The sea was dark. Somewhere a generator coughed. "If this were only about war," Ernst said, "I would leave." Karl looked at him. "Would you?" Ernst did not answer. "That is the problem with saying only," Karl said. "There is never only. There is the war. There is the mathematics. There is the dream. There is the uniform. There are the prisoners none of us are supposed to discuss. There is London. There is the Moon. There is what we tell ourselves so we can sleep." Ernst kicked at the sand. "You make it sound impossible to be clean." "Perhaps it is." "Then why stay?" Karl looked toward the test stands. "Because I want to know if a machine can leave the Earth." "And if it lands on people?" Karl had no answer that did not shame him.

After the war, the rocket men moved into a new story with astonishing speed. Von Braun and many of his colleagues surrendered to American forces and were brought to the United States. The Soviet Union took other German engineers, parts, documents, and lessons. The war had ended, but the rocket had not become innocent. It had changed employers. In the emerging Cold War, rockets meant missiles, deterrence, national prestige, satellites, and eventually astronauts. The same control problem remained: how to make a tall, explosive, accelerating body go where humans intended.

In Huntsville, Alabama, years later, an American engineer named Joe Miller found himself sitting across from von Braun in a meeting about a launch vehicle. Joe is fictional, but the room is assembled from many such rooms: slide rules, ashtrays, drawings, schedules, German accents, Southern heat, military urgency, and the new language of space. "We can make the structure lighter," Joe said, tapping the drawing. Von Braun looked over his glasses. "Everything can be made lighter once." "Once?" "Before it fails." A few men laughed. Joe did not. He had learned that von Braun could be charming and ruthless in the same minute. "Guidance margin is still tight," another engineer said. "Guidance margin is always tight," von Braun replied. "That is why we have guidance." The joke was thin, but the truth was not. To reach orbit, a rocket must not merely rise. It must pitch, roll, accelerate, stage, steer, and arrive with enough horizontal speed to keep falling around Earth. A cannonball needs only a barrel. A satellite launcher needs a nervous system. It needs instruments that say where it is pointing, computers or analog devices that decide what correction is needed, actuators that move engines or fins, and a structure that survives being corrected.

In October 1957, the Soviet Union launched Sputnik. The small beeping sphere did more to American confidence than many larger weapons. It passed overhead, invisible to most eyes but present in imagination. The United States suddenly felt not only behind but observed. Joe heard the news on a car radio. His wife, Margaret, turned down the volume after the announcer repeated it for the third time. "Is it bad?" she asked. "It is brilliant," Joe said. "That is not what I asked." "It is bad because it is brilliant." She looked through the windshield at the dark road. "Can we do it?" Joe thought of von Braun's team, of Redstone and Jupiter-C, of arguments over which American program should launch first, of politics wrapped around engineering until no one could tell where one ended. "Yes," he said. "If they let the right people try."

On January 31, 1958, Explorer 1 reached orbit on a Juno I rocket derived from Jupiter-C, developed under von Braun's Army Ballistic Missile Agency team. At the press conference, von Braun stood with William Pickering and James Van Allen, holding up a model while cameras captured relief disguised as triumph. The satellite discovered what would be called the Van Allen radiation belts. The United States had answered Sputnik, not fully, not finally, but enough to breathe. In a living room in Ohio, a boy named Peter watched the newsreel with his father. "Is that the German rocket man?" Peter asked. His father, who had served in Europe, did not answer immediately. "Yes." "Is he ours now?" The father looked at the smiling men on the screen. "That is a complicated word." "Ours?" "Now." The Space Race transformed von Braun's public image. He appeared on television explaining spaceflight to Americans. He became a salesman for the future, a man who could make orbital mechanics seem like a national adventure. Children knew his name as a name pointing upward. Yet the shadow of Peenemünde did not vanish. It moved with him, sometimes ignored, sometimes challenged, always present for those willing to look.

At NASA's Marshall Space Flight Center, the dream grew huge. Saturn V was not only a rocket; it was a controlled collapse of impossibilities. Its first stage burned kerosene and liquid oxygen through five F-1 engines, producing a force almost too large for imagination. Its upper stages used liquid hydrogen and oxygen. Its guidance system had to keep the stack stable as it rose, pitched, accelerated, shed mass, separated stages, and inserted spacecraft toward the Moon. If Peenemünde had taught that a rocket must listen quickly, Apollo demanded that it listen perfectly enough to carry human beings.

Joe stood near a test stand years later when an F-1 engine fired. The sound was not sound at first. It was a physical fact. Beside him, a younger engineer named Ruth Harris pressed her hands over her ears even through protection. "People are going to sit on top of five of those," she shouted. "Three people," Joe shouted back. "That does not improve it." "No." "Do you ever think about where this began?" Joe kept his eyes on the flame. "Which beginning? Goddard? Peenemünde? Sputnik? Kennedy's speech? The budget office?" "You know which one I mean." He did. Everyone did, though not everyone said it. "Yes," he said. "I think about it." "And?" "And I still want the Moon landing to work." Ruth nodded slowly. "So do I. That is what bothers me."

On July 16, 1969, Saturn V lifted Apollo 11 from Earth. The rocket did what all rockets must do: it balanced on fire. It corrected, listened, adjusted, and survived its own power long enough to send human beings toward another world. Millions watched. Some saw American victory over the Soviet Union. Some saw the fulfillment of ancient longing. Some saw engineering at its most magnificent. Some saw a former German weapons engineer standing near the summit of American achievement and could not separate wonder from unease. In Peter's house, now older, he watched with his father again. Peter asked, "Still complicated?" His father looked at the Saturn V climbing on television, the picture ghostly and delayed. "More," he said. "But magnificent." "Yes." "Can both be true?" His father leaned back. "They usually are." That may be the most honest sentence about rockets. They are magnificent and compromised. They expand human possibility and carry human politics. They are born from dreams and budgets, curiosity and fear, prestige and military need. Control engineering does not decide the reason for launch. It makes launch possible. It gives ambition a steering system.

Today, rockets rise for many reasons. Some carry astronauts to the International Space Station. Some deploy weather satellites that warn people before storms. Some carry communication satellites that connect ships, villages, aircraft, banks, soldiers, scientists, and families. Some carry telescopes that look backward in time. Some are military. Some are commercial. Some are national flags in another form. Some are billionaire visions, some planetary science, some climate monitoring, some navigation, some surveillance. The old mixture remains: wonder, profit, security, science, pride.

At a modern launch site, a young engineer named Amara Singh watches a reusable booster descend toward a landing pad. She is fictional, but the scene is familiar now: a rocket coming back upright, engine relit, legs deployed, trying to do in reverse what earlier generations barely dared to do upward. Beside her stands Leo, a systems engineer who has seen enough launches to distrust applause before shutdown. "It still looks fake," Amara says. "It is the opposite of fake," Leo says. "It is very expensive honesty." The booster falls tail-first, engine plume bright against the afternoon. "Landing burn start," a voice says over the loop. Amara stops breathing. Rockets make people do that. The booster corrects sideways, rights itself, slows, drops, hovers in the mind though not in physics, and touches down. Applause breaks out. Leo waits another second. "Now," he says, and claps. "You waited." "Rockets deserve manners. Do not congratulate them early." Amara laughs, but her eyes stay on the vehicle. "Why do we still do this?" "Because Earth needs satellites." "That is the practical answer." "Because nations like flags." "That is the cynical answer." "Because people with money like impossible projects." "That is the annoying answer." Leo looked toward the sky where the upper stage had disappeared. "Because the sky keeps asking questions." That one satisfied her less, but stayed longer.

Why explore space now? Not because space is empty romance. It is not. Space is infrastructure. It is climate observation, GPS timing, disaster monitoring, communications, science, defense, weather prediction, agriculture, ocean tracking, and planetary defense. It is also the Moon again, Mars in imagination, asteroids, telescopes, and the human refusal to let the horizon remain only symbolic. Some reasons are noble. Some are commercial. Some are military. Some are childish in the best sense. Some are childish in the worst.

But none of it happens without control. A rocket must hold attitude during ascent. A spacecraft must orient solar panels, point antennas, fire thrusters, dock, land, avoid tumbling, manage fuel, survive delays, and sometimes make decisions far from immediate human command. The farther from Earth we go, the more control becomes patience built into machines. Mars cannot be landed on by joystick from Earth; the delay is too long. A rover must enter, descend, and land through a sequence of controlled events that cannot wait for a human hand. The dream of exploration becomes real only when correction can happen in time.

At the launch site, Amara and Leo walk back after the landing. "Do you ever think about von Braun?" she asks. "Yes." "How do you hold it?" Leo takes a while before answering. "Carefully." "That is not enough." "No. But it is where I start. The history of rockets is not clean. War pushed them. Forced labor built them. Fear funded them. Competition accelerated them. Science ennobled them. Television romanticized them. Engineers made them work. All of that is true." "And we still launch." "Yes." "Why?" He stops near a fence where the recovered booster stands cooling in the distance. "Because refusing to use knowledge does not erase where it came from. But using it without memory repeats the worst part of the bargain." Amara looks at the rocket. "So memory is part of control?" "It had better be."

The rocket chapter, then, is not a simple ascent from weapon to Moon to reusable booster. It is a story of continuity and transformation. The V-2 showed that a guided rocket could reach beyond artillery and strike a city. Explorer 1 showed that a modified military rocket could place a scientific instrument in orbit. Saturn V showed that controlled fire could carry humans to another world. Modern rockets show that control can even bring a booster back to land, turning what was once discarded into something like an aircraft with no wings. The same central problem remains through all of it: power must be guided. Fire must be made obedient. A rocket without control is only an explosion with ambition. A rocket with control becomes a weapon, a satellite launcher, a Moon vehicle, a telescope carrier, a rescue supply ship, a commercial service, a national symbol, a scientific instrument. That is why rockets are morally revealing. Control gives them capability, but purpose gives them meaning. The equations do not decide whether a rocket carries explosives to a city or instruments to study the Van Allen belts. The gyroscopes do not decide whether the launch is part of war, prestige, commerce, or curiosity. The guidance system only asks: where have you told me to go? Karl at Peenemünde, Joe in Huntsville, Ruth at the test stand, Peter before the television, Amara at the landing pad: each stands at a different point in the same flame-lit history. Each sees something different. War. Opportunity. Unease. Pride. Wonder. Responsibility. The rocket rises through all of them. And every time it does, it must solve the same problem in the first seconds. Do not tip. Do not tumble. Listen faster than the human eye. Correct before error becomes destiny. Balance on fire, and go where we asked.

Chapter 9

The Sky Becomes a Battlefield

In war, the way is to avoid what is strong and to strike at what is weak.

—Sun Tzu, *The Art of War*, Chapter 6, trans. Lionel Giles

In 1915, near a muddy road behind the Western Front, Private Thomas Bellamy heard an engine before he saw the aircraft, and the sound disturbed him more than shellfire because it did not belong to the war he had been trained to understand. Shells came from somewhere, even if the somewhere was hidden. Rifle fire came from a line. Cavalry, when it still appeared, came over ground. But the airplane came from above, small and insolent against the clouds, looking almost harmless until someone shouted that it was spotting for artillery. Thomas was fictional, but the fear was not. Around him men looked upward with the awkwardness of creatures whose eyes had not evolved for battle in the sky. A sergeant named Wilkes raised his rifle and fired. Others followed, not because they expected to hit the aircraft but because doing nothing felt like consent. At a nearby battery, Captain Henri Leclerc, an officer modeled on those who improvised the first air defenses, watched the little machine circle. "It is not here to kill us," Thomas said when he reached the gun position with a message.

Leclerc did not lower his binoculars. “That is the old mistake.” “It carries no bomb.” “It carries eyes.” A French gunner laughed nervously. “Then we must shoot out its eyes.” Leclerc pointed toward a field gun that had been dragged onto a crude mounting to fire at a steep angle. “We must first teach a cannon to look up.”

The first great difficulty of anti-aircraft defense was that the target had escaped the plane of ordinary war. A soldier could aim a rifle at a man because both shared the ground. A gun crew could range a trench because the trench was fixed. An airplane moved in three dimensions, changed speed, changed height, and turned the sky into a volume instead of a ceiling. The old habits of sight and aim became inadequate. A shell fired at an aircraft had to arrive not where the aircraft was, but where it would be, and it had to burst near enough that fragments might do what direct impact rarely could. That meant estimating range, altitude, speed, direction, wind, fuse time, gun laying, shell flight, and the delay between command and explosion. Control entered as prediction under panic. Leclerc tried to explain this to Thomas while the gunners elevated the piece. "If I aim at it, I miss behind it. If I aim ahead, I must know how far ahead. If I set the fuse wrong, the shell bursts under it or after it. If I wait for certainty, it has already told their artillery where we are." Thomas looked up again and felt the new helplessness of modern war. "So the gun must shoot into the future." "All guns do," Leclerc said. "Aircraft merely make the future embarrassingly visible."

By the end of the First World War, armies had learned that the airplane was not a curiosity but a system of reconnaissance, artillery correction, bombing, and psychological pressure. The word "Archie," used by British soldiers for anti-aircraft fire, carried comedy because soldiers often protect themselves with jokes before they have protection in steel. But behind the jokes came instruments: height finders, sound locators, searchlights, predictors, better fuses, dedicated anti-aircraft guns, and organizations that treated air defense as a discipline rather than a desperate upward gesture. Imagine Leclerc years later meeting a young engineer named Martin Ellis, at a postwar demonstration of a mechanical predictor. He turns cranks while a model aircraft moves across a plotting board. "You give me bearing, elevation, and range," he says, "and the machine estimates where the target will be when the shell arrives." Leclerc watches the gears with suspicion. "The machine guesses?" "It calculates." "A polite guess." "A disciplined one." He smiles at that, because discipline is the bridge between soldier and engineer. In anti-aircraft fire, the controller is not a single device but a chain of observation, estimation, computation, command, actuation, and explosion. If any part lags, the aircraft flies through the gap.

Between the wars and during the Second World War, the sky filled not only with single aircraft but with formations, raids, bombers, dive bombers, torpedo planes, and the terrible arithmetic of mass attack. Cities became targets from above, ships became targets from above, and soldiers learned that an aircraft could be overhead before the body had time to choose fear properly. At Bofors in

Sweden, a company better known before the war for heavy guns and steel made one of the most important anti-aircraft weapons of the twentieth century: the 40 mm Bofors gun, especially the L/60 model that entered service in the 1930s and spread through Allied and other forces. Let us place a Swedish engineer, Nils Dahl, in a workshop at Karlskoga, speaking with a naval officer named Commander Erik Lindholm as the gun cycles through a test. "It must hit aircraft," Lindholm says. Nils wipes oil from his hand. "No, Commander. First it must keep firing." "That is not enough." "It is the beginning of enough. A gun that jams is an opinion. A gun that feeds, trains, elevates, and repeats can become defense." The Bofors 40 mm mattered because it sat in a practical middle ground: heavier and more lethal than machine guns, faster-firing and more flexible than large high-altitude guns, useful against low-flying aircraft that arrived quickly and left little time for calculation. Its success was not a single miracle of barrel or shell. It was an integration problem: ammunition feed, recoil, training speed, sights, crew drill, mount stability, manufacturing quality, and eventually radar direction and proximity fusing around larger air-defense systems.

On a destroyer in the Pacific, a U.S. Navy gun captain named Joe Ramirez stands behind a Bofors mount while a Japanese aircraft drops toward the ship. Beside him is Chief Petty Officer Samuel Greene, who has learned to speak in commands because sentences waste time. The ocean is bright, the aircraft is dark, and the ship is turning hard enough that Joe's stomach lags behind his feet. "Lead him!" Greene shouts. Joe is already trying. The sight demands that he imagine the aircraft not as a thing but as a moving solution. The mount turns. The barrels bark. Empty casings clatter. "More right!" someone yells. "He's crossing!" The difficulty for the sailor is not merely courage. It is cognition under acceleration. The target grows, the ship rolls, smoke hides the sight picture, other guns fire, orders collide, and the aircraft becomes both a machine and an intention. A Bofors crew had to become a human servomechanism: one man tracking, another loading, another clearing, another commanding, all trying to close the loop between eye, hand, gun, shell, and target. "Did we get him?" Joe asks after the aircraft splashes short. Greene does not answer immediately. He is already searching the sky for the next one. "You do not get the sky," he says.

"You borrow seconds from it."

Radar changed air defense because it gave the defenders something soldiers had wanted since Thomas first heard the engine in 1915: warning before sight. During the Second World War, radar networks, command centers, fighter direction, searchlights, guns, and eventually proximity fuses turned air defense into a layered control system. The British Chain Home radar network did not shoot down aircraft by itself, but it changed the timing of decisions during the Battle of Britain. Information arrived early enough for commanders to choose where to send fighters, when to scramble, and how to conserve strength. In a operations room, a Women's Auxiliary Air Force plotter named Elsie Moore moves markers across a table while Air Chief Marshal Hugh Dowding, real head of RAF Fighter Command, watches reports accumulate. "If the reports are wrong," Elsie says quietly, "the pilots go to empty sky." Dowding answers, "If the reports are late, they go to burning airfields." The same principle applied to guns. Radar could measure range and bearing; predictors could compute lead; servos could slew mounts; proximity fuses could make a shell explode when near a target instead of requiring a perfect timed burst. The loop tightened. The defender's problem was still to predict the future, but the instruments had become faster than unaided human judgment.

The arrival of missiles after the Second World War changed the problem again. A jet bomber flew faster than a propeller aircraft. A ballistic missile did not fly like an airplane at all. It rose, left the lower atmosphere, and returned at speeds that compressed decision time into a moral and technical emergency. Anti-aircraft guns could not simply be enlarged into answers. Surface-to-air missiles such as the Soviet S-75, known to NATO as the SA-2, and American systems such as Nike Ajax and later Nike Hercules moved air defense from gun laying to guided interception. Now the defender launched a machine that had to chase another machine. Guidance became central: command guidance from the ground, radar tracking, semi-active homing, active seekers, fusing, and control surfaces that corrected the missile's path faster than any gun crew could move steel. Imagine a Soviet engineer, Sergei Lebedev, speaking with a young radar operator after an S-75 battery drill in the late 1950s. "The missile is a gun barrel that leaves the ground," the operator says. Sergei shakes his head. "No. A gun barrel is finished aiming when the shell departs. A missile continues to argue after launch." "With whom?" "With the target, with the radar, with the air, with its own errors." This is the essence of missile defense against aircraft: the controller is now airborne, and interception is a moving conversation between sensors, computers, propulsion, and fins.

When the target became another missile, the conversation became harsher. To defend against a ballistic missile is to hit, or explode near, an object moving extremely fast, often with little warning, sometimes accompanied by decoys, and often carrying a warhead whose arrival would be catastrophic. Cold War missile defense was therefore never merely an engineering problem; it was strategy, deterrence, politics, and fear made mathematical. In the United States, programs from Nike Zeus to Safeguard tried to answer whether a country could defend cities or missile fields from nuclear attack, while in the Soviet Union similar questions shaped Moscow's defenses. Let us imagine an American engineer named Ruth Kaplan, borrowed in name from another chapter's grid engineer but now older and working as a consultant, speaking in 1968 with Colonel James Alden at a test range. "You want certainty," she says. "I want survival," Alden replies. "That is a more emotional word for certainty." He points toward the radar dishes. "Can the system tell a warhead from junk?" Ruth does not answer quickly. "Sometimes." "That is not enough." "It may be all physics and budget permit." The challenge of missile defense is that control must operate at the edge of knowledge. Sensors see fragments, heat, radar returns, trajectories. Computers estimate. Interceptors launch. Corrections happen in seconds. The defender is trying to create a controlled meeting between two objects that would otherwise never politely meet.

Late in the twentieth century and into the twenty-first, air defense became layered because no single weapon could answer every threat. A large radar-guided missile might reach a high aircraft but be wasteful against a cheap drone. A gun might defeat a low target but not a ballistic missile. A fighter might intercept a bomber but not every cruise missile. A system such as Patriot, first designed for aircraft and later adapted for missile defense, belongs to a family of defenses in which radars, command posts, launchers, interceptors, datalinks, and crews form a distributed controller. Other systems fill other layers: short-range guns, shoulder-fired missiles, medium-range missiles, naval systems, electronic warfare, decoys, and passive defense. In an imagined training room, a Patriot engineer named Daniel Weiss speaks with a air-defense officer, Captain Noor Haddad. "People call it a missile battery," Noor says. "They picture tubes." Daniel taps the radar display. "The tubes are the last sentence. The radar is the first word." "And the computer?" "Grammar." "And the crew?" Daniel looks around the room at the operators learning to trust and distrust their screens at the same time. "The crew decides whether the sentence should be spoken." This is important because modern air defense is not automatic in the childish sense. It is automated under human rules, constrained by identification, rules of engagement, civilian risk, ammunition scarcity, and the terror of mistaking a friendly aircraft for an enemy.

The current war between Russia and Ukraine has shown the old problem in a new swarm of forms. Since Russia's full-scale invasion of Ukraine on February 24, 2022, Ukrainian soldiers and civilians have faced aircraft, helicopters, cruise missiles, ballistic missiles, glide bombs, Shahed-type one-way attack drones, reconnaissance drones, loitering munitions, and small quadcopters that drop grenades or correct artillery fire. Russia has also had to defend against Ukrainian missiles and drones reaching ships, airfields, depots, refineries, and cities. The sky is no longer only high altitude. It is layered from the stratosphere to the height of a rooftop. A Ukrainian soldier named Oksana Melnyk stands near a mobile air-defense team at night, listening for a Shahed drone whose engine sounds like a machine refusing to be ashamed. Beside her is Mykola, older, a former mechanic. "Too slow for a missile," he says. "Too cheap for what we fire at it," Oksana answers. Their pickup carries a heavy machine gun and a tablet receiving warnings from a wider network. Somewhere far away, a radar has seen something. Somewhere closer, civilians have heard it. Searchlights move. Phones buzz. "There," Mykola says. The drone is a dark speck with a sound larger than its body. The challenge is brutally practical: a million-dollar interceptor cannot be the answer to every cheap drone, but a cheap drone can still destroy a transformer, an apartment block, a fuel tank, or a radar. Control here includes economics. The loop must not only be fast and accurate; it must be affordable enough to repeat all night.

Drones have changed the soldier's experience of the sky because they make surveillance intimate. In 1915 Thomas feared an airplane that could direct artillery. In Ukraine, a soldier in a trench may fear a quadcopter hovering close enough to watch him light a cigarette. The aircraft is no longer necessarily a roaring machine flown by a pilot risking his own body. It may be a small electric device controlled from kilometers away, or an autonomous one-way munition following coordinates, or a first-person-view drone diving through a window. The air-defense engineer therefore faces targets that are slow, small, numerous, low, and cheap, exactly the combination that embarrasses systems built for fast, large, expensive threats. Imagine a Ukrainian electronics volunteer named Danylo meeting a Swedish engineer named Karin Dahl, Nils's granddaughter, who has come to advise on integrating radar cues with guns. "Your grandfather's Bofors was built for airplanes," Danylo says. Karin corrects him gently. "My grandmother's generation built for what frightened them." "Now this frightens us." He holds up a small drone propeller recovered from a field. Karin turns it in her hand. "Then the question is the same: how quickly can we see it, decide it is hostile, point something at it, and do so without spending the country empty?" Danylo nods. "And how do we do that when there are twenty?" Karin looks toward the workshop where soldiers are modifying mounts, tablets, radios, and sensors. "Then it is no longer a duel. It is traffic control under fire."

The answer emerging in Ukraine and elsewhere is not one weapon but a layered, adaptive, improvised control system. Old guns return with new sights. Gepard self-propelled anti-aircraft guns, once considered Cold War equipment, have proved valuable against drones and low-flying threats because radar-directed cannon fire can be cheaper than missiles and fast enough for close defense. Bofors guns, ZU-23s, heavy machine guns, MANPADS, Patriots, S-300s, NASAMS, IRIS-T, electronic jammers, acoustic sensors, phone apps, spotter networks, and decoys all occupy pieces of the same defensive ecology. Some systems kill. Some confuse. Some warn. Some cause the attacker to waste weapons. This is control as architecture rather than gadget. A commander must decide what to engage, with what, at what range, and what to save for the next wave. In a Ukrainian command post, Colonel Iryna Shevchenko stands before a map while an American adviser named Mark Ellis and a German technician named Lena Vogt argue over interceptors. "If we fire Patriot at every drone, we lose the battery without losing the warhead," Iryna says. Mark points to a track. "That one may be masking something faster behind it." Lena adds, "Or it is bait." Iryna closes her eyes for half a second, the only luxury command gives her. "Then we classify, layer, and hold discipline." The phrase sounds calm. It is not calm. It is the control problem of scarcity under attack.

Protection against missiles and drones also includes protecting people before interception succeeds or fails. Sirens, shelters, dispersal, camouflage, hardened substations, rapid repair crews, blackout discipline, deception, and redundancy are part of air defense because no shield is perfect. The soldier's burden is to engage the sky; the civilian's burden is to live under the possibility that engagement will fail. In Kyiv, Kharkiv, Odesa, Belgorod, Sevastopol, and many other places touched by the war's reach, people have learned the grammar of alerts: ballistic threat, cruise missile threat, drones in the region, air defense active, stay in shelter. A schoolteacher named Halyna asks Oksana after a night attack, "Did you shoot it down?" Oksana answers, "One of them." "There were more?" "There are always more." Halyna looks toward children sleeping in coats in a corridor. "Then what is victory?" Oksana thinks of radar tracks, ammunition counts, generators repaired at dawn, a drone burning in a field, a missile that did not hit the hospital, and the sound of another alarm before breakfast. "Sometimes victory is a thing arriving nowhere important." That is not a slogan generals prefer, but it is often the truth of defense. Success is an absence: no crater, no fire, no funeral, no headline large enough to notice.

From the first improvised guns of the First World War to the Bofors 40 mm, from radar-directed batteries to surface-to-air missiles, from Cold War missile shields to Ukrainian mobile teams hunting drones at night, the central control problem has remained recognizable while the targets have changed. The defender must sense, identify, predict, decide, aim, launch or fire, observe the result, and prepare for the next threat. Each era has shortened the time available and multiplied the ambiguity. The airplane made soldiers look up. The bomber made cities look up. The missile made computers look up. The drone has made everyone look up, including the soldier in a trench and the child in a subway station. Air defense is therefore not only the story of weapons. It is the story of human attention stretched into machines: binoculars into radar, shouted bearing into data link, handwheel into servo, gun sight into predictor, local alarm into national network. Thomas with his rifle, Ingrid with the Bofors mechanism, Joe behind the naval mount, Elsie at the plotting table, Sergei beside the missile rail, Noor in the Patriot training room, Oksana under the drone track: all of them live inside the same hard sentence. The sky is no longer empty. If danger can come from above, protection must learn to rise faster than fear, calculate faster than sight, and correct before impact becomes history.

CHAPTER 10

Cleaner Air, Lower Energy

Nature does nothing uselessly.

—Aristotle, *Politics*, Book I, Chapter 2

On a hot afternoon in Los Angeles in the late 1960s, a mother named Elena Morales stood at a bus stop on Wilshire Boulevard with one hand over her son's mouth, not because he had said anything rude, but because the air itself felt rude. The boy, Mateo, was nine years old and angry that adults kept inventing reasons for him not to run. His chest had tightened during recess. The school nurse had told him to sit still. Now his eyes watered and the mountains had disappeared behind a brown veil. "Where did they go?" he asked, pointing toward the place where the San Gabriels should have been. "The mountains?" Elena said. "They are still there." "Then why can't we see them?" She looked at the traffic crawling past, chrome bumpers, gasoline engines, tailpipes idling at every light, and she thought of all the machines that had promised California freedom. Cars had given families beaches, suburbs, jobs, dates, deliveries, Sunday drives, and the feeling that the horizon belonged to anyone with a full tank. They had also given the city a chemistry experiment large enough to breathe. "Because," she said, "we have made the sky dirty." Mateo considered this with the literal severity of a child. "Then someone should clean it." A man in a short-sleeved shirt beside them laughed without happiness. "Good luck telling Detroit that." Elena looked down the boulevard where sunlight had become yellow and hard. "Maybe California will."

California did. The state's struggle with smog made it one of the most important laboratories in environmental control. Los Angeles had a special cruelty: sunlight, geography, temperature inversions, millions of vehicles, refineries, industry, and a basin that could hold pollution like a lid over a pot. By the middle of the twentieth century, people could see and feel that prosperity had a tailpipe. The response was not one invention but a political and technical feedback loop. Citizens coughed, doctors warned, scientists measured hydrocarbons, nitrogen oxides, carbon monoxide, ozone, and particulates, regulators set limits, engineers redesigned engines and exhaust systems, monitors measured again, standards tightened, and industry protested, adapted, and improved. The California Air Resources Board, created in 1967, became a central actor in this loop. California pushed vehicle emission standards before the rest of the United States moved as strongly, and because the California market was too large to ignore, its rules shaped cars sold far beyond the state. In an imagined hearing in Sacramento, a young regulator named Ruth Alvarez listens as an automobile executive named Mr. Hanley warns that the proposed standards are unrealistic. "You are asking us to change the automobile," he says. Ruth answers, "No. The air is asking you. We are only writing down its complaint."

The early fight was often presented as cleanliness against performance, regulation against freedom, air against the automobile. That was not entirely false. Emission controls cost money, added complexity, and sometimes caused badly tuned engines to drive poorly. But the deeper story is more interesting for control engineering because cleaner engines eventually required more precise engines. The catalytic converter, introduced widely in the 1970s, could transform carbon monoxide, unburned hydrocarbons, and nitrogen oxides into less harmful gases, but it worked best only when the engine fed it exhaust of the right chemical character. Too much fuel and the mixture ran rich, wasting gasoline and producing carbon monoxide and hydrocarbons. Too little fuel and the mixture ran lean, raising other problems and making nitrogen oxides harder to control. The oxygen sensor in the exhaust gave the engine a way to listen to its own combustion. Electronic fuel injection and engine control units then adjusted the air-fuel mixture many times per second. The car became a rolling feedback system: measure oxygen, infer mixture, adjust fuel, observe exhaust, correct again. What began as a demand to reduce pollution helped push the automobile from carburetors and mechanical compromise toward electronic control, sensors, actuators, diagnostics, and software.

The small part that made this possible deserves more attention because it is one of the purest examples of feedback hidden in ordinary life. Bosch developed the lambda sensor, or oxygen sensor, from ceramic oxygen-measurement work that had roots in high-temperature industrial sensing. In the accounts often repeated in automotive histories, the late-1960s Bosch development work was carried out under the supervision of Dr. Günter Baumann, whose team turned a zirconium-dioxide ceramic element, coated with porous platinum electrodes, into a sensor tough enough to sit in hot exhaust and compare the oxygen in the exhaust stream with the oxygen in outside air. That comparison produced an electrical signal. The signal told the fuel system whether combustion had been rich or lean. The word lambda, written as λ, named the air-fuel equivalence ratio; the magic point was near $\lambda = 1$, the stoichiometric mixture where gasoline and air are balanced so that a three-way catalyst can simultaneously attack hydrocarbons, carbon monoxide, and nitrogen oxides. The engineering difficulty was that the useful region was narrow. The catalyst did not need a driver with a good conscience. It needed a controller with a fast ear.

Volvo then made the idea real in production. In 1976, for the 1977 model year and especially for the demanding U.S. and California market, Volvo introduced its Lambda Sond system on the 240/260 family. The Swedish spelling, sond, means probe, and the name was well chosen: the car had acquired a chemical sense organ. Stephen Wallman at Volvo was one of the key figures behind the system, and the 1977 SAE paper by Grunde Engh and Stephen Wallman described the arrangement with admirable control-engineering clarity: three essential new components, an exhaust-gas composition sensor, an additional feedback loop to the continuous fuel-injection system, and the catalyst. The breakthrough was not simply that Volvo bolted on a cleaner exhaust box. It closed the loop around combustion. In an imagined late-night test cell in Göteborg, Wallman stands with a Bosch engineer beside a Volvo B21 engine running on a dynamometer. The exhaust analyzer prints its slow judgment while the lambda signal flickers faster than either man can read by eye. "The catalyst wants balance," the Bosch engineer says. Wallman nods. "Then we must stop asking the carburetor to guess." "And if the sensor drifts?" "Then the law will find us." "The law?" Wallman points toward the exhaust pipe. "No. That." Outside the laboratory, California's air was the customer, and it was no longer accepting promises.

In a garage in Pasadena in 1983, an older mechanic named Frank DeLuca watches a younger technician named Nina Chen plug a diagnostic tool into a sedan. Frank grew up with carburetors, timing lights, vacuum hoses, and the belief that an engine revealed itself through sound, smell, and a hand on the fender. Nina has learned those things too, but she also trusts voltage. "It is running rich," she says. Frank sniffs the exhaust. "I could have told you that." "The oxygen sensor told me first." "Sensors don't know engines." Nina smiles. "Engines don't know laws." Frank wipes his hands on a rag. "Laws don't know customers." She points to the analyzer. "Customers know when they fail the smog test." The exchange is comic only because both are right. Regulation changed the customer's problem. The car was no longer judged only by whether it started, accelerated, and got the family to work. It had to pass an invisible test. Its waste became measurable. Once waste was measurable, it could become controllable.

This is the first lesson of pollution control: many forms of environmental harm are errors that were once unmeasured. A tailpipe without a sensor is a confession without a listener. A smokestack without monitoring is a promise made in the dark. A building without an energy meter is a household that knows it is poor only at the end of the month. Control begins when invisible flows become signals. In Los Angeles, smog made pollution visible in the most brutal way, but modern environmental control depends on seeing more than the eye can see. Lambda sensors in cars, continuous emissions monitors in power plants, particulate sensors, building energy dashboards, smart thermostats, variable-frequency drives, grid frequency measurements, satellite observations, and air-quality stations all turn waste into feedback. Elena at the bus stop could see brown air, but she could not see ozone chemistry, grams per mile, combustion temperature, or evaporative emissions from fuel tanks. Engineers and regulators had to give society instruments before society could correct itself.

The second lesson is that reducing pollution and reducing energy consumption often support each other, but not automatically. Burning less fuel usually means producing less carbon dioxide and often fewer pollutants. A more efficient engine wastes less heat. A well-insulated building needs less heating and cooling. A motor driven only as fast as necessary consumes less electricity. A ship that trims speed can save fuel and emissions. But pollution chemistry can be stubborn. An engine can be fuel efficient and still produce nitrogen oxides if combustion temperature is high. A diesel engine can use less fuel than a gasoline engine and still create particulate and NOx problems without aftertreatment. A factory can save energy while shifting pollution elsewhere. Control engineering lives in these tradeoffs. It does not say, "Optimize one number and the world is healed." It asks which outputs matter, which constraints must not be violated, and what price is being hidden. In a California laboratory, imagine Ruth Alvarez years later watching a test vehicle on a dynamometer with an engineer named Samir Patel. "Fuel economy improved," Samir says. "Good," Ruth answers. "NOx?" He hesitates. "Higher under some conditions." "Then the car has not solved the problem. It has moved the problem." Samir sighs. "You regulators never let us celebrate." "I will celebrate when the lungs and the fuel tank both win."

The history of cars after California's early regulations shows how often both could win. Electronic engine management improved drivability, emissions, and fuel use because it could adapt in ways fixed mechanical systems could not. A cold engine needs one strategy; a warm engine needs another. Acceleration, cruising, idling, altitude, fuel quality, catalyst temperature, and aging sensors all change the control problem. The engine controller became a small, fast bureaucracy under the hood, constantly negotiating among torque, emissions, fuel economy, temperature, knock, catalyst protection, and driver demand. Later hybrid vehicles made the control problem larger and more elegant. A Toyota Prius or similar hybrid is not merely a gasoline car with a battery attached.

It is an energy manager. It decides when to run the engine, when to shut it off, when to use the electric motor, when to charge the battery, when to recover braking energy, and how to keep the engine operating in efficient regions. Regenerative braking is feedback with memory: motion that would have become heat in brake pads becomes electrical energy stored for later use. In an imagined taxi in Los Angeles, Mateo, now grown and asthmatic but stubbornly alive, rides with a driver named Grace who points proudly at the display. "See that? When I brake, it gives me energy back." Mateo laughs. "When I was a boy, cars gave us smog back." Grace says, "This one still burns gas." "I know. But it wastes less of its apology."

Buildings offer another field where control can make pollution and energy visible at once. A badly controlled building heats rooms nobody occupies, cools air and then reheats it, runs fans at full speed because nobody taught them partial effort, leaves lights burning for absent people, and fights itself through poor scheduling. The occupants experience this as ordinary discomfort: one office too cold, another too hot, conference rooms stale, windows open above radiators, a thermostat turned to an extreme because the previous hour was unpleasant. The energy system experiences it as waste. Modern building control uses sensors, occupancy schedules, variable air volume boxes, heat pumps, chilled water loops, CO_2 sensors, weather forecasts, and optimization to keep people comfortable with less energy. In a university building in Copenhagen, a facilities engineer named Mikkel walks with Professor Anna Berg, who is angry because her office is cold. "The new system is supposed to save energy," she says, wrapping her scarf tighter. "It is saving too much in my office." Mikkel checks a tablet. "The occupancy sensor thinks you are absent." "I am a professor. Sitting still is part of the job." He laughs and changes a parameter. "Then the building has mistaken thought for vacancy." The joke carries a warning. Energy control that ignores humans becomes tyranny by thermostat. Good control saves energy by understanding use, not by making people miserable enough to surrender.

Factories and process plants reveal the same principle at larger scale. A refinery, paper mill, cement kiln, steel plant, wastewater facility, or chemical reactor is a world of flows, temperatures, pressures, reactions, motors, pumps, compressors, valves, and emissions. In older plants, operators often ran equipment conservatively: extra air, extra pressure, extra pumping, extra heating, extra margin. Margin protects production, but it can also become waste fossilized as habit. Advanced process control can reduce energy use by keeping operation closer to optimal conditions while respecting safety and quality limits. A furnace with better oxygen control burns fuel more completely. A compressor with variable-speed control avoids throttling losses. A distillation column with model predictive control can meet product purity without excessive reflux. A wastewater plant can aerate only as much as biology requires rather than blowing air continuously into tanks like a nervous giant. In an imagined control room at a Danish food-processing plant, an operator named Jens watches a trend line flatten after a new controller is commissioned. "So the pump slows down when demand falls?" he asks. The young control engineer, Leila, nods. "It follows the process instead of bullying it." Jens studies the quieter motor. "For twenty years we ran it full speed and strangled it with a valve." "That worked." "So does driving with one foot on the accelerator and one on the brake."

Electric grids are perhaps the grandest example because every watt saved or shifted can change what must be generated. Control in the grid used to mean, above all, matching generation to demand so frequency stayed stable. It still means that, but the environmental meaning has deepened. Wind and solar power vary with weather; demand varies with human life; batteries, electric vehicles, heat pumps, industrial loads, and buildings can respond if the control architecture allows them to. A smart thermostat that precools a house before a peak, a charger that waits until wind power is plentiful, a battery that discharges during a dirty peak hour, a factory that shifts a flexible process by thirty minutes, and a grid operator that dispatches resources with carbon intensity in mind are all parts of a cleaner energy loop. But here too the human dialogue matters. In a California neighborhood during a heat wave, a utility sends a message asking customers to reduce power use between 5 and 9 p.m. Elena's grandson, Lucas, now an engineer, watches his smart thermostat raise the set point. His daughter Sofia complains, "The house is thinking for us again." Lucas says, "The grid is under stress." "The grid is not doing homework in this heat." "Neither are you, apparently." She rolls her eyes. He softens. "If millions of houses do a little, we may avoid turning on the dirtiest plants." Sofia looks at the thermostat. "So being slightly uncomfortable is climate policy?" "Sometimes. But the system should ask fairly, pay fairly, and never ask the same people to suffer every time." Demand response can reduce energy peaks and pollution, but if designed badly it becomes another way to export discomfort to those with less power.

Control can also expose a paradox: efficiency can make a service cheaper, and cheaper service can increase use. Economists call this rebound. A more efficient car costs less to drive, so people may drive more. Better lighting reduces electricity per lumen, so cities may illuminate more surfaces. A data center can improve cooling efficiency and still consume more total energy because computation grows. This does not mean efficiency is useless. It means efficiency is not morality by itself. A controller that reduces waste inside one device must be placed inside a larger social controller that asks how much activity is enough, who benefits, and what total emissions result. Frank the mechanic, now retired, might put it more bluntly while watching traffic on a cleaned-up but still crowded freeway. "We made each car cleaner," he says to Nina. "Then we filled every lane with them." Nina answers, "Would you prefer the old smog?" "No." "Then what?" He looks toward the downtown skyline, visible now on many days his youth would have hidden it. "I prefer remembering that cleaner is not the same as clean, and efficient is not the same as wise."

The California story therefore should not be told as a simple victory in which regulation solved the car. Los Angeles air became dramatically cleaner than in the worst smog decades, and vehicle emission controls were a major reason, along with fuel changes, industrial controls, monitoring, and public policy. Yet the region still struggles with ozone, particulates, freight corridors, ports, wildfire smoke, and the unequal geography of exposure. Children near freeways and warehouses may breathe more harm than children in wealthier hills. Cleaner technology can reduce average pollution while injustice remains concentrated. Control engineers know about averages and peaks. A system that looks acceptable on average may still fail where the load is highest. Environmental control must therefore ask not only how many tons are reduced, but where the remaining tons go and whose lungs serve as the buffer.

At the end of the chapter, imagine Elena Morales very old, sitting on a bench with Mateo and his granddaughter Sofia at Griffith Observatory after a rain has washed the basin clean. The city lies below them in impossible detail: freeways, towers, neighborhoods, ports, aircraft, reservoirs, stadiums, roofs glittering with solar panels, cars moving like metallic blood through concrete arteries. The mountains are visible. Elena points to them with a hand made thin by age. "When your grandfather was a boy, some days those were gone." Sofia frowns. "Gone?" Mateo says, "Hidden. By us." A plane descends toward LAX. A bus climbs the hill. Somewhere below, millions of controllers are working without ceremony: engine computers trimming fuel, building systems slowing fans, inverters matching sunlight to the grid, chargers waiting, treatment plants aerating, monitors sampling air, thermostats bargaining with heat. Sofia asks, "Did control fix it?" Mateo looks at the city for a long time. "Control helped us stop wasting without noticing. It helped us make dirt visible. It helped machines apologize less with smoke." Elena adds, "But people had to demand it." The child considers this. "So clean air is a machine?" Elena shakes her head. "No. Clean air is an agreement that uses machines." That is the most important sentence. Control can reduce pollutants and energy consumption when society gives it the right measurements, the right limits, and the courage to correct profitable waste. It can make a car burn cleaner, a building breathe smarter, a plant run with less excess, and a grid use cleaner power. But the set point is not chosen by the sensor. The set point is chosen by people deciding what kind of air they are willing to breathe, what kind of energy they are willing to waste, and what kind of future they refuse to poison for convenience.

Chapter II

Cruise Control, Traffic, and the Illusion of Smooth Motion

Our life is frittered away by detail. Simplify, simplify.

—Henry David Thoreau, *Walden*, "Where I Lived, and What I Lived For"

The Los Angeles freeway offered David Morales a brief and misleading kindness as he guided his old silver sedan onto the 101 near Hollywood on a Thursday afternoon. The lanes were full but moving, the sun was low enough to flash between buildings, and the familiar green signs promised destinations with the calm authority of objects that never have to arrive anywhere themselves. Beside him sat his sister Elena, visiting from Sacramento, holding a paper cup of coffee with both hands because she did not trust either the lid or Los Angeles. In the back seat, David's teenage son Nico wore headphones and looked out at the river of cars as if it were a boring documentary about adult failure. "Not bad," Elena said. David smiled, because every Angeleno knows that saying not bad on the 101 is a kind of prayer. "Wait until Cahuenga." "Is that a place or a warning?" "Both." The sedan settled into the lane. David reached to the steering wheel,

pressed a button, and the word CRUISE appeared in small green letters on the dashboard. He lifted his foot slightly from the accelerator. The car kept going. "There," he said. "Modern civilization." Elena looked at the dashboard. "Your car is from 2009." "Civilization ages unevenly."

For several minutes, the machine did exactly what it promised. David had set the speed at fifty-eight miles per hour, and the car treated that number as a commandment. When the road rose gently, the controller opened the throttle enough to keep speed from falling. When the road dipped, it eased back. It did not know Los Angeles, or appointments, or impatience, or whether the driver was tired. It knew a reference value and a measured value. It compared the desired speed with the actual speed, detected error, and moved the throttle to reduce that error. If wind, slope, or engine load disturbed the car, the loop corrected. "So it is like the thermostat," Nico said, one headphone lifted. David glanced in the mirror. "You were listening?" "Against my will." Elena smiled. "Yes, like the thermostat, except the room is a freeway and your father is pretending he is relaxed." David tapped the steering wheel. "I am relaxed." A motorcycle split lanes beside them, loud enough to erase the sentence. "Deeply relaxed," Elena said.

Cruise control is one of the gentlest introductions to feedback because its ambition is modest. It does not decide where to go. It does not steer. It does not know whether the driver should have left earlier. It only tries to keep speed near the driver's chosen set point. The sensor measures vehicle speed. The controller compares that speed with the set point. The actuator adjusts engine torque. Disturbances arrive as hills, wind, road surface, passengers, air conditioning load, and the small mechanical moods of an aging car. The driver remains responsible for the larger world: lanes, exits, curves, police cars, rain, debris, and above all other vehicles. In light traffic this division feels elegant. The driver chooses a speed; the machine handles the tedious correction. Human attention is freed from the ankle. But the bargain is fragile because a freeway is not a test track. The car can hold speed only while holding speed is the right thing to do. As traffic thickens, the set point becomes less like a desire and more like a fantasy.

The first brake lights appeared near Universal City. They did not appear all at once. They glowed in small red clusters far ahead, then vanished, then returned closer, like an animal breathing under the pavement. David's sedan continued at fifty-eight because that was what he had asked it to do. The distance to the car ahead began to shrink. Elena lifted her coffee. "David." "I see it." "The car sees it?" "No." "That seems like an important limitation." He touched the brake, and the cruise control disengaged with a small click that felt almost offended. The sedan slowed. A newer SUV in the next lane, glossy and expensive, eased down more gracefully. Its brake lights came on without drama. Then it accelerated again, then slowed again, keeping a measured gap from the vehicle ahead. Elena nodded toward it. "That one knows distance." David watched the SUV. "Adaptive cruise control. Radar or camera or both. It follows the car in front." "Why don't you have that?" Nico asked. "Because I bought your orthodontics instead." "My teeth do not maintain following distance." "No, but they cost like they should."

Adaptive cruise control changes the problem from speed regulation to spacing regulation. Instead of merely asking, "Am I going the speed the driver chose?" the car also asks, "How far am I from the vehicle ahead, and how quickly is that distance changing?" It uses radar, camera, lidar in some systems, or sensor fusion to estimate range and relative speed. Then it adjusts throttle and braking to maintain a time gap: perhaps one, two, or three seconds of following distance. In smooth traffic, this feels almost magical. The car accelerates when the lane opens, slows when the car ahead slows, and relieves the driver of the exhausting work of tiny speed corrections. Yet it is still a local controller. It sees the vehicle ahead, maybe a few vehicles indirectly through motion, but it does not know the entire freeway. It does not know that a delivery truck three miles ahead changed lanes awkwardly, that an on-ramp is feeding too many cars into too little space, that a nervous driver tapped the brakes once and created a disturbance that will travel backward through traffic for ten minutes. It is smarter than old cruise control, but it is still mostly a neighborly intelligence. It watches the back of the car in front and tries to behave.

By 4:46 the 101 had become the 101. Speed fell, rose, fell again. The sedan crept past a billboard advertising a streaming series about beautiful people with terrible judgment. David accelerated to thirty-two, then braked to seventeen, then rolled at nine, then accelerated again because the lane ahead opened, then braked harder because the opening had been a rumor. Around them, thousands of drivers repeated the same ritual with different degrees of annoyance and skill. "Why does it do this?" Elena asked. "There is no accident. No construction. Nobody selling fruit in the lane. It just pulses." Nico leaned forward. "It is like a wave." "Exactly," David said. "A wave made of bad decisions." Elena looked at him. "All traffic explanations eventually become moral criticism." "Fine. A wave made of delayed corrections." That was better. A traffic jam is often not a solid object but a moving pattern. A driver brakes slightly. The driver behind reacts a little later and brakes a little more. The next driver reacts later still. The disturbance travels backward even while every car is trying to move forward. From above, traffic can look like compression waves in a fluid: dense regions and loose regions, stop and go, acceleration and braking, energy wasted as heat in brake pads and anger in human bodies.

The strange part is that no driver needs to intend the jam. Each driver can be reasonable locally and still produce disorder collectively. If the car ahead slows, you slow. If a gap opens, you accelerate. If someone merges, you make room or refuse to make room and force someone else to correct. Each action makes sense from inside one windshield. But traffic is a coupled system. The output of one driver becomes the input to the next. Human reaction time adds delay. Human impatience adds gain. Limited visibility hides the state of the system. Different cars accelerate and brake differently. Different drivers choose different gaps. Some stare at phones. Some fear being cut off and close space aggressively. Some brake with a feather; others brake as if killing a snake. The result is a decentralized control system with thousands of controllers, each pursuing a private objective: keep my speed, protect my gap, reach my exit, do not let that person in, do not miss the meeting, do not be the fool who leaves too much space. The freeway has no single mind, only a crowd of small minds connected by brake lights.

"If everyone just drove twenty-five steadily, would it clear?" Elena asked after they had moved one car length in thirty seconds. David laughed, then stopped because she was serious. "Maybe, if everyone actually did it." "Why not put up a sign? Twenty-five. Stay steady." Nico removed his headphones entirely. "Because people are people." "That is not an engineering answer." "It might be the engineering answer," David said. A constant slow speed can carry more flow than stop-and-go motion if it prevents hard braking and preserves stable gaps. The problem is not that steady motion is physically impossible. The problem is that it is socially difficult when every driver controls their own vehicle according to local perception and private incentives. If David leaves a large smooth gap, another driver may merge into it. If he refuses to accelerate when space opens, the driver behind may honk or change lanes. If he tries to damp the wave by driving gently, he pays a private cost for a public benefit. Traffic stability asks individuals to behave as if they are part of a system. Rush hour teaches them to behave as if they are defending territory.

This is where centralized or coordinated control becomes tempting. Ramp meters already do a simple version of it by controlling how quickly cars enter the freeway. Variable speed limits can slow traffic upstream before it reaches a bottleneck, reducing shock waves. Navigation systems can distribute drivers across routes, though sometimes they merely move congestion into neighborhoods never designed to absorb it. Connected vehicles could share information beyond line of sight: the car ahead is braking because five cars ahead someone cut in; slow gently now. Autonomous fleets could coordinate acceleration and spacing with machine patience. In theory, a central traffic controller with enough sensors, communication, and authority could ask many vehicles to move at a steady speed lower than what impatient individuals would choose, producing a smoother, safer, more efficient flow for everyone. The freeway would become less like a crowd and more like a conveyor belt with choices.

But that phrase choices is where the argument begins. Elena imagined the system aloud. "So the city tells your car: you may go twenty-three now, not thirty-seven." "The city already tells me speed limits." "Speed limits are boundaries. This sounds like command." "Maybe advisory command." "That is a phrase invented to lose a public hearing." Nico grinned. "I would watch that hearing." David pictured it: commuters at microphones, civil libertarians, traffic engineers, delivery companies, parents, police, software vendors, people who loved trains, people who hated trains, people who considered lane-changing a constitutional right. The technical case for coordination can be strong. Smooth traffic reduces fuel consumption, emissions, crashes, stress, and wasted time. The political case is harder because cars are not only transportation devices. They are private rooms, status objects, escape machines, and symbols of self-direction. To tell people that their cars should obey a central smoothing algorithm is to touch a nerve much older than traffic engineering. "My car, my choice," Elena said. "Even when your choice creates my brake lights," David answered. "Especially then," she said.

Control always raises the question of who chooses the set point. In the old cruise-control sedan, David chose fifty-eight, and the machine obeyed until reality forced his foot back into the loop. In adaptive cruise control, the driver chooses a desired speed and following distance, while the car modifies speed to respect the vehicle ahead. In coordinated traffic control, the set point could come from a system optimizing flow across many people at once. That may be better engineering and worse politics, or better politics if people value clean air and shorter commutes enough, or worse ethics if the system privileges some routes, neighborhoods, or vehicles over others. A central controller can smooth waves, but it can also encode priorities: freight over commuters, wealthy corridors over poor ones, main roads over side streets, data-rich drivers over invisible pedestrians. Decentralized freedom can produce jams. Centralized optimization can produce obedience. Neither word, freedom or optimization, is innocent.

Near downtown, traffic opened suddenly, as if the jam had been a dream the freeway no longer wished to discuss. David accelerated, then stopped himself from resuming cruise control. The memory of the brake lights remained in his foot. The adaptive SUV beside them moved ahead with polished calm, keeping distance, never looking embarrassed. Elena watched it go. "Would you want one?" she asked. "The newer car? Yes." "The central traffic system?" He took longer to answer. "Some days, yes. Some days I want everyone, including me, forced to behave intelligently." "And other days?" "Other days I do not want a transportation department, a car company, and a software update deciding how urgent my life is." Nico said from the back, "Your life is not urgent. You are going to buy groceries." "Groceries can be urgent." "Not the way you shop." Elena laughed, and the car moved with the lane, another private capsule inside the public machine.

At the exit, David looked back at the freeway from the ramp. The cars below were no longer individual stories from that distance. They were dots in a stream, red lights flickering, gaps opening and closing, a pattern no one driver had chosen and every driver had helped create. Cruise control had promised him smooth motion by controlling one car. Adaptive cruise control promised smoother motion by controlling one car relative to another. Traffic itself asked for something larger: coordination across many cars, many ramps, many intentions. The lesson of the 101 at rush hour is not that drivers are foolish, though some are, or that machines are wise, though some are useful. The lesson is that local control has limits. A thousand reasonable corrections can become a wave. A thousand private freedoms can become a public delay. To smooth the freeway, someone or something must see more than one bumper ahead. The difficult question is whether we are willing to be seen that way, and whether the controller that sees us will serve the road, the city, or itself.

Chapter 12

Self-Driving Cars and the Ethics of Correction

Wisely and slow; they stumble that run fast.
—William Shakespeare, *Romeo and Juliet*, Act II, Scene iii

On an October night in San Francisco, Nina Park waited at the corner of Fifth and Market and watched three kinds of traffic argue with one another. There were cars with human drivers, tired and impatient under the city lights. There were buses breathing at the curb. There were cyclists threading between lanes as if the rules were suggestions negotiated in motion. And there were the driverless cars, white and quiet, carrying no one or someone, it was sometimes hard to tell, moving with the careful confidence of machines that had been taught the city in numbers. Nina was a paramedic, but the street was real. She was off duty, holding a paper bag of takeout, thinking about whether the noodles would survive the walk home. Across the intersection, a Cruise robotaxi paused with the unnatural patience of a vehicle that did not mind being stared at.

"There goes one," said Marco, the friend beside her. "One what?" "The future." Nina looked at the Cruise vehicle. "The future has its hazard lights on too often." "You don't like them." "I don't dislike them. I dislike not knowing what they understand." That was the unease many people in San Francisco had learned to name in fragments. The robotaxis were impressive, annoying, careful, clumsy, safer than humans in some moments, stranger than humans in others. They stopped for things human drivers ignored and ignored things human beings would have found socially obvious. They became part of the city before the city had agreed on what kind of part they were. Some riders loved them. Some cyclists hated them. Firefighters complained about them blocking emergency scenes. Engineers defended them with data. Supervisors demanded caution. Tourists photographed them. Residents learned to walk around them.

Marco watched the Cruise car begin to move. "At least it doesn't get drunk." "Neither does a concrete wall," Nina said. "That doesn't make it a driver." A few blocks away, in a Cruise operations room, a fictional remote assistance worker named Leah watched a set of vehicle tiles on a screen. The room itself belongs to the real architecture of autonomy: maps, alerts, support channels, logs, dashboards, and humans watching fleets of vehicles whose decisions were mostly made elsewhere, inside software moving through a city faster than any supervisor could narrate. She did not drive the cars. That was the point. But she watched for conditions that escaped the neat categories of training data and road rules.

"Panini is downtown," said Omar, another operator. "Panini?" Leah asked. "Vehicle name. Don't blame me." "I blame whoever named a robotaxi after lunch." Omar laughed. "Quiet night?" Leah knew better than to answer that. In operations rooms, calling a night quiet is a way of insulting probability. The first impact did not involve the Cruise car. A human-driven Nissan struck a pedestrian in the crosswalk and threw her into the adjacent lane, into the path of the Cruise autonomous vehicle. That sentence is easy to write and almost impossible to inhabit. A person was walking through a city. A driver hit her. Her body became an object in another lane before any machine or human in that lane had enough time to make the world fair again. The Cruise vehicle braked, but it struck her. Then it stopped. Then, in the part of the event that would change the company, the vehicle attempted a pullover maneuver and dragged the woman about twenty feet.

On the corner, Nina heard the sound before she understood the sequence. "That was a person," she said. Marco froze. "What?" "Call 911." "Nina–" "Now." She was already moving. The paper bag fell to the sidewalk. The city contracted into training: traffic, victim, vehicle, danger, bystanders, airway, bleeding, fire, police, space. People shouted with the terrible redundancy of crowds. "She was hit!" "The car stopped!" "Don't move her!" "It's moving!" That last shout was the one Nina would remember. It had a different shape from the others. It was not only fear that someone had been struck. It was fear that the machine had not understood the moral status of the moment. A vehicle in an accident scene is no longer merely a vehicle in traffic. It becomes part of a rescue scene. Its first obligation is not to clear a lane. Its first obligation is not to behave like traffic is still the main problem. Its first obligation is to stop being dangerous.

The Cruise car's logic had another idea. After detecting a collision, the system attempted to move to the side of the road, a maneuver intended in ordinary circumstances to reduce traffic risk. But this was not an ordinary circumstance. A human being was underneath or entangled with the vehicle. The correction made sense in one frame and became horrifying in another. Nina reached the scene with another bystander, a man in a denim jacket who kept saying, "I saw it, I saw it," without yet explaining what he had seen. The Cruise vehicle had stopped. The pedestrian was trapped. The street had become a room full of people who wanted to help and could not agree how.

"Back up," Nina said. "Give her air. Give us space." "Are you a doctor?" someone asked. "Paramedic." "Can we lift it?" "Not yet. Wait for fire." Marco arrived, phone to his ear. "They're coming." "Tell them vehicle versus pedestrian, trapped under autonomous vehicle, Fifth and Market." He repeated the words. Autonomous vehicle made people look at the car again, as if the phrase itself had weight. The car sat with its lights and sensors, silent, no driver to shout at, no face to accuse, no one behind the windshield to say I didn't see her or I panicked or I am sorry. That absence changed the anger. People are used to anger having a target.

In the operations room, alarms changed color. "Panini collision event," Omar said. Leah leaned forward. "Show me." The screen gave fragments: vehicle status, location, sensor feeds, system messages. It did not give the street's human temperature. It did not show the way bystanders' voices changed when they realized the vehicle had moved after the first stop. It did not show the moral meaning of twenty feet. "Did it reposition?" Leah asked. Omar's face tightened. "Looks like a pullover maneuver after collision detection." "After hitting a pedestrian?" "I don't know what it classified." That was the sentence. I don't know what it classified. A human on the street saw a person under a car. A system may have seen an impact, an object, a stopped state, a traffic hazard, a need to exit the lane. Between those descriptions lay the entire problem of autonomous driving. To control a car, a machine must not only steer and brake. It must understand what kind of world it is in. It must know when the rules of driving have given way to the rules of rescue.

"Freeze the logs," Leah said. "Already happening." "Escalate." "Already happening." "Then escalate louder." The real aftermath did not belong only to emergency responders. It moved quickly into regulators' offices, company meetings, newsrooms, city politics, federal investigations, and public trust. The California Department of Motor Vehicles suspended Cruise's autonomous deployment and driverless testing permits on October 24, 2023, stating that Cruise's vehicles were not safe for public operation and that the company had misrepresented safety information. The California Public Utilities Commission followed by suspending Cruise's authority to provide driverless passenger service. NHTSA opened an investigation into Cruise pedestrian-safety incidents. Cruise later recalled 950 robotaxis for a software update related to collision response, and the company admitted to filing a false report with NHTSA. Kyle Vogt, Cruise's co-founder and CEO, resigned in November 2023.

But before the regulatory language, there was a person in the street. The pedestrian's name was not part of the public story in the way corporate names were. That is proper. Privacy is one of the few protections left after a body becomes news. But the anonymity can make the event too abstract. A woman was gravely injured. Her body was struck first by a human-driven car and then by a driverless car. The second vehicle did not intend harm. That matters technically and not at all physically. Harm does not require intention. A week later, in a Cruise conference room, a safety engineer named Daniel Cho sat across from a manager named Priya Shah. On the wall was a timeline of the crash. Daniel had slept badly since watching the full sequence. Priya had slept badly for different reasons: regulators, press, subpoenas, employees, families, the sudden sense that the company had crossed from ambition into accusation.

"The first car caused the initial impact," Priya said. Daniel looked at her. "Yes." "We need to be precise about that." "We also dragged her." Silence. "The vehicle executed a post-collision pullover behavior," Priya said. Daniel closed his eyes. "That sentence is why people hate us." "It is accurate." "It is incomplete in the worst possible direction." Priya folded her hands. "Then say it your way." "The car did not understand that the collision scene had changed the meaning of movement. It treated clearing the lane as safety. In that situation, safety meant do not move." "And why did it not understand?" Daniel pointed to the diagram. "Because our system was built around categories that failed. It lost or mishandled the pedestrian state after impact. It did not maintain the right memory of what it had just run over. It did not know enough to treat the vehicle itself as a hazard to a trapped person. It corrected one problem by creating another."

Priya looked tired. "Control failure." "World-model failure. Classification failure. Memory failure. Safety-case failure. And yes, control failure. The wheels moved because the system believed moving was the right control action." That is why this belongs in a book about control. A self-driving car is not only a steering controller. It is perception, prediction, planning, decision-making, actuation, remote assistance, fleet operations, company policy, regulatory permission, and public trust, all compressed into a machine sharing streets with people who did not consent individually to be part of the experiment. The public sees the car turn left or stop at a crosswalk. The real control problem is deeper: what does the car think is happening, how sure is it, what does it remember, what is it allowed to do when the world becomes strange, and who is responsible when its correction is wrong?

A human driver can also fail terribly. The first vehicle in the October crash was driven by a person, and that person fled. Any honest discussion must keep that fact in view. Humans speed, drink, text, rage, misjudge, panic, fall asleep, and flee. The promise of autonomous vehicles has always depended on this indictment of human driving: machines do not get drunk, do not text, do not show off, do not fall asleep, do not carry grudges from work into the accelerator pedal. If they can see and decide well enough, perhaps they can reduce the enormous harm caused by human drivers.

But the Cruise crash showed the other side. Machines do not panic, but they also do not understand unless understanding has been built, tested, and bounded. They do not flee in shame, but they may execute a maneuver that no ashamed human would attempt because the human would hear the screams. They do not get distracted, but they may fail to preserve the one fact that suddenly matters most. They do not intend harm, but they can still produce it with mechanical calm. At a public hearing, let us place a San Francisco resident named Rosa Morales at the microphone. She is a teacher, not an engineer, and she has waited two hours to speak for one minute.

"I am not against technology," she says. "My son uses a wheelchair. Technology gives him things my grandmother would have called miracles. But when your cars block fire trucks, stop in intersections, or hurt someone and then move like the street is still normal, you are asking us to trust a system we cannot question." A commissioner asks, "What would trust require?" Rosa looks toward the company representatives. "A car that knows when to stop. A company that tells the whole truth. And rules that do not wait for a disaster before becoming serious." That last demand is not anti-technology. It is exactly what safety-critical control requires. The system must know its limits. The company must know its obligations. The regulator must know enough to ask hard questions before the public learns through injury. A self-driving car is not safe because it completes thousands of rides. It is safe only if the rare, ugly, edge cases have been imagined deeply enough. Pedestrians thrown by another car. Fire hoses in the street. Police gestures. Construction workers. Fallen cyclists. Fogged sensors. Ambiguous debris. A person under the vehicle. These are not exceptions to the driving world. They are part of it.

Months later, after layoffs, resignations, investigations, and software changes, Daniel meets Leah in a cafe far from the office. Both have left Cruise in this version, though many real people stayed, left, argued, grieved, defended, or rebuilt in their own ways. The company they joined had promised a safer future. The company they left had become a warning. "Do you still believe in self-driving cars?" Leah asks. Daniel watches a human driver roll through a stop sign outside. "Yes," he says. "That was not a confident yes." "It is a damaged yes." "What does that mean?" "It means I still think human driving is terrible. I still think automation can help. I still think a car that never drinks, never texts, and sees in all directions could save lives. But I no longer trust the phrase safer than human as if it answers everything."

Leah nods. "Safer for whom, in what scenario, with what fallback, under what oversight?" "Exactly." "You sound like a regulator." "I sound like someone who watched a car make a sentence out of a woman." Leah looks down at her coffee. "Don't say it that way." "Why?" "Because it is too close." He does not apologize. She does not ask him to. The ethics of correction is not a slogan. It is the problem at the center of autonomy. Every controller acts because it believes the world would be better after the action. Brake. Swerve. Yield. Proceed. Pull over. Stop. Call for help. In ordinary driving, these actions are small. In extraordinary moments, they become moral. The same pullover behavior that might reduce risk after a fender-bender can become catastrophic when a person is trapped. The same rule that says clear the lane can become wrong when the lane has become a rescue scene.

Control engineering often speaks of objectives. Minimize risk. Maintain traffic flow. Avoid collision. Obey rules. Preserve passenger comfort. Reach destination. But the world does not always present objectives in a neat order. Sometimes the system must recognize that one objective has become sacred and all others must fall away. Do not drag the person. Do not move into the firefighter. Do not block the ambulance. Do not treat the abnormal scene as a normal road with inconvenience added. The October 2023 Cruise crash did not prove that autonomous vehicles can never be safe. It proved something narrower and more important: that safety is not demonstrated by ordinary competence alone. A city is made of edge cases. A safety case must include the moments when classification fails, when memory must persist, when the system is damaged or confused, when the morally correct action is to do less rather than more. Control must sometimes mean restraint so absolute that the machine refuses its own next instruction.

At the end of the chapter, imagine Nina walking again through Fifth and Market months later. The street is ordinary because streets are cruelly good at becoming ordinary again. People cross with coffee. Buses exhale. A cyclist curses at a taxi. A driverless car from another company passes slowly, sensors turning, passenger seat empty. Marco, beside her, says, "Still the future?" Nina watches the car stop for a pedestrian. "Maybe," she says. "That is all you give it?" "Maybe is generous." "What would make it yes?" She waits until the pedestrian reaches the curb and the car moves again. "When it understands that stopping is not a pause in driving," she says. "Sometimes stopping is the whole duty."

Chapter 13

The Elevator, the Crane, and the City in Motion

A great city is not to be confounded with a populous one.

—Aristotle, *Politics*, Book VII, Chapter 4

Rain had followed Mara Stein into the lobby of a forty-seven-story Chicago office tower, where the architect stood watching people disappear into polished steel boxes. The lobby smelled of raincoats, coffee, stone, perfume, and the faint electrical cleanliness of a building waking itself for work. A young analyst named Peter had one eye on his phone and one eye on the elevator indicators. A courier balanced a tube of drawings under his arm. Two lawyers continued a conversation that had clearly begun in a taxi and would probably survive until lunch. Nobody looked afraid. That was the remarkable thing. They were about to step into a suspended room, trust cables, brakes, motors, doors, sensors, software, counterweights, inspection regimes, and invisible calculations, and be lifted hundreds of feet above the street. Yet the emotion in the lobby was not awe. It was impatience. "Why is it taking so long?" Peter muttered. Mara smiled.

Her colleague Daniel Price, a structural engineer, heard him and whispered, "The true sign of civilization is complaining that a miracle is thirty seconds late." Mara looked at the elevator doors. "No. The true sign is forgetting it is a miracle at all."

An elevator is a control system that has succeeded so completely that most people experience it as furniture. You press a button; a car arrives; doors open; people enter; doors close; motion begins; motion ends; doors open again at the correct floor. Each of those verbs hides a condition. The car must know where it is in the shaft. It must accelerate without throwing passengers off balance, travel without exceeding safe speed, decelerate smoothly, stop level with the floor, keep doors closed while moving, keep the car still while people cross the threshold, refuse to move when overloaded, avoid collisions with other equipment, respond to emergency calls, and fail in ways that make rescue possible. The old romance of machines often celebrates power: the engine, the turbine, the crane, the rocket. The elevator teaches a quieter lesson. Power becomes useful only when obedience is exact. The motor must not merely lift. It must lift when permitted, stop when required, and wait when the door lock says no.

Mara had brought Daniel to the building because she was designing a tower and had begun to fear the elevators more than the wind. "The columns I understand," she said as they entered a car. "The bracing I understand. Even the dampers have a kind of honesty. But these shafts eat the building from the inside." Daniel pressed thirty-two. "Elevators always eat the building." "That sounds like a curse." "It is geometry. People want to go up, and every vertical path you reserve for movement is floor area you cannot rent, live in, or fill with sunlight." The doors closed. The car rose with a force so well shaped that the body noticed mostly by losing the lobby. "Without elevators," Daniel said, "a skyscraper is a stack of inaccessible promises." "And with too many elevators?" "A skyscraper becomes a transportation machine with offices attached." Mara watched the floor numbers climb. The elevator was not an accessory to height. It was the permission for height. A building could be designed upward only if people could be moved upward fast enough, safely enough, and without surrendering most of the plan to shafts.

Frank Lloyd Wright understood this problem in the extravagant way only a visionary could. In 1956 he unveiled his proposal for The Illinois, also called the Mile-High Illinois, a 528-story skyscraper imagined for Chicago, one mile high and meant to hold a population so large it resembled an urban district turned vertical. The drawings were magnificent, implausible, seductive, and impossible in the practical terms that make buildings cruel to dreams. The elevator problem was central. To move tens of thousands of people through a mile-high tower using ordinary mid-century elevators would require so many shafts, so many transfers, and so much waiting that the building would begin to devour itself. Wright therefore imagined technology beyond what existed: high-speed, atomic-powered, gyroscopic elevators that could move at a mile a minute. Let us place Wright in an imagined conversation with a skeptical elevator engineer named Ruth Kaplan after a lecture. "Mr. Wright," Ruth says, "your building is taller than our elevator technology." Wright smiles as if this is not an objection but evidence that the technology is underambitious. "Then the elevator must grow up." "A shaft is not a hallway. Cars cannot pass through one another because architecture is inspired." "Then invent a better shaft." Ruth looks at the mile-high drawing. "You have designed a city in the air." "Yes." "Then you must also design its subway."

That is the hidden truth of tall buildings. Above a certain height, elevator design becomes urban planning. The lobby is a station. The shaft is a track. The car is a train with a very short route and a vertical timetable. Passengers arrive in waves: morning up-peak, lunch traffic, down-peak, interfloor travel, hotel check-ins, conference surges, evacuations, maintenance trips, deliveries, children pressing buttons, executives demanding express service, cleaners moving carts at inconvenient hours. A single elevator can be understood as a controlled machine. A group of elevators must be understood as a controlled society. Which car should answer which call? Should one car wait for a likely crowd, or leave now with fewer people? Should cars be zoned by floor? Should passengers enter destinations before boarding so that the system can group them intelligently? Should one elevator skip a nearby call to prevent worse delay later? These are not questions of muscle but of dispatch, prediction, and fairness.

Marja-Liisa Siikonen's work at KONE belongs to this less visible side of elevator control. She studied elevator traffic, group control, simulation, call allocation, and what KONE later called People Flow: the planning of how people actually move through buildings, not how architects wish they moved on clean drawings. In one of her papers on elevator traffic simulation, waiting time and ride time become measures of service, while passenger traffic patterns and call allocation algorithms determine whether a bank of elevators feels intelligent or irritating. This may sound dry until you stand in a lobby with two hundred people at 8:58. Then it becomes social reality. Imagine Siikonen explaining this to Mara at a conference in Helsinki. "The elevator is not only a car," she says. "It is a promise about waiting." Mara asks, "A promise to whom?" "To everyone, which is why it is difficult. If the system serves one person immediately, it may delay thirty others. If it optimizes the group, one person may feel ignored." "So elevator control is politics." Siikonen smiles. "In a small room, with doors."

Destination control made this politics more explicit. In older systems, passengers entered a car and then declared their floor by pressing buttons inside. The controller learned destinations only after the group had already formed. With destination dispatch, passengers enter their desired floor in the lobby, and the system assigns them to a car before boarding. The advantage is that people going to nearby floors can be grouped, stops can be reduced, round trips can become shorter, and handling capacity can improve. The disadvantage is psychological. The old elevator let people feel as if they chose inside the car. The new one tells them where to stand before the doors open. In Mara's Chicago lobby, Peter approaches a destination panel and presses 32. The screen says C. He walks toward car C, then stops. "How does it know?" he asks Daniel. "You told it." "I told the panel. I did not tell that car." "The panel and the car speak." Peter looks offended. "I miss buttons." Mara laughs. "You miss the illusion that the button was listening only to you."

Safety remains the deeper discipline beneath this convenience. Elevator control is not simply about reducing waiting time. It is about forbidding catastrophe. Doors must not close with unacceptable force on a body. A car must not move with doors open. Brakes must hold. Overspeed governors must act. Buffers must absorb. Interlocks must prevent elegant software from doing physically stupid things. Emergency systems must behave when power fails, fire alarms sound, water enters pits, earthquakes shake guide rails, or passengers panic. The safest command is often refusal. Do not move. Do not open. Do not answer that call yet. Do not trust the sensor if another sensor disagrees. The elevator is polite because it is constrained. It serves people by obeying limits that passengers rarely see and would not want to negotiate.

The city above the lobby depends on the same idea. Cranes move loads through air that people occupy below. Escalators coordinate steps, motors, comb plates, handrails, skirts, emergency stops, and crowds who do not look down. Automated train doors must align with platforms, refuse to close on bodies, and communicate with signaling systems that govern motion beyond the station. A construction hoist lifting workers outside a tower, a hospital elevator prioritizing a patient on a stretcher, a goods lift carrying food to a restaurant kitchen, a subway escalator emptying commuters into daylight: all are examples of controlled motion near human bodies. Urban life is full of machines that must be strong without being rough, fast without being reckless, automatic without being arrogant. The city works because motion is sequenced. The crane does not swing whenever it wants. The train does not open doors between stations. The elevator does not treat the shaft as open road.

And yet the old shaft still limits the future. Ropes, counterweights, and one-car-per-shaft logic impose architectural consequences. Double-deck elevators, sky lobbies, express zones, destination control, lighter ropes, better motors, and more sophisticated group control all try to make vertical transportation less hungry. KONE's UltraRope, for example, addressed the weight problem of very tall elevator travel by replacing conventional steel hoisting ropes with a lighter carbon-fiber-based technology, making new heights more practical. But an even more radical idea returns to Wright's complaint: what if the elevator were not trapped in one vertical line? What if it could behave less like a bucket in a well and more like a train in a network?

The newest dream is no longer atomic-powered, but it is still a challenge to the old shaft. TK Elevator's MULTI system uses linear motors and ropeless cars so that multiple cabins can move in a loop, vertically and horizontally, changing shafts rather than merely reversing direction in one hoistway. The metaphor is not a single elevator but a metro inside a building. Several cars can occupy a circulation system, one following another, with switching that allows horizontal movement between vertical paths. This is exactly the kind of idea Wright reached toward without the technology to implement it: not just faster cars, but a different topology of movement. In an imagined test tower, Mara watches a cabin move sideways and says, "That feels illegal." Daniel says, "Only because elevators trained you to think in one dimension." "If cars can change shafts, the building becomes a track diagram." "Yes. And the control problem becomes much harder." Cars must be routed, separated, scheduled, switched, accelerated, stopped, evacuated, and trusted. A ropeless multidirectional elevator is freedom only if control becomes more disciplined than ever.

At the end of the day, Mara returns to the Chicago lobby. The morning impatience is gone; the evening impatience has replaced it. People descend from offices, rise to meetings, cross paths with cleaners, couriers, tourists, and security guards. The elevators keep opening their doors like small negotiations with gravity. Peter emerges from car C and says to no one in particular, "Finally." Mara almost tells him that he has just ridden inside one of civilization's great control achievements, that his complaint is a tribute, that boredom is what safety feels like after engineering has done its work. Instead she lets him go. The doors close. Another car arrives. Somewhere above, motors pull, brakes wait, sensors count, algorithms assign, passengers gather, and a tower remains usable because motion has been made orderly. The elevator's lesson is not merely that control can move us. It is that control can make dangerous motion socially ordinary. A city becomes vertical only when people trust a hidden system enough to step inside, turn their backs to the doors, and think not of falling, but of being late.

Chapter 14

Smart Buildings and the Politics of Comfort

We shape our buildings; thereafter they shape us.

—Winston Churchill, speech to the House of Commons, 28 October 1943

Snow followed Nora Lind into the glass lobby of the Nordhavn Civic Center in Copenhagen, where she stamped her boots onto a mat placed by someone who understood both weather and litigation. The building opened around her with the confidence of expensive intelligence: turnstiles recognized badges, elevators assigned cars before people pressed buttons, lights rose gently in corridors that had been dim a moment earlier, and warm air descended from slots in the ceiling as if the building had noticed her discomfort and decided to be kind. Nora worked for the municipality's public health office on the sixth floor. She had not come to think about control engineering. She had come to argue about budgets, answer emails, and find out why the conference room called Birch had been freezing every Tuesday for three weeks. At the security desk, Henrik looked up from his screen. "Morning." "Is Birch still an ice cave?" Nora asked. Henrik smiled without joy. "The system says Birch is comfortable." "Then the system should attend the meeting."

The civic center had been advertised as a smart building, though smart was one of those words that grew less precise the more money stood behind it. The building knew when people entered, where they gathered, which rooms were booked, how much carbon dioxide accumulated in the air, how much sunlight struck the south facade, how much electricity cost in the next hour, whether the district-heating supply was under strain, whether a filter was clogging, whether a pump was drawing more current than expected, whether a window contact had been opened, whether a toilet was leaking, whether an elevator car was idling in the wrong zone, and whether the cafeteria refrigerators had drifted above their limit. It did not know that Nora always felt cold after cycling through sleet, that Malik in finance wore shirtsleeves in February, that the west meeting rooms smelled faintly of carpet glue when ventilation was reduced, or that people became less charitable when they had been shivering for forty minutes under a vent. A smart building is intelligent in the way many control systems are intelligent: it sees what has been instrumented, optimizes what has been defined, and mistakes the rest for noise.

In the basement plant room, where the intelligence became pipes, valves, dampers, pumps, coils, and humming cabinets, a building engineer named Søren Madsen watched trends on a monitor while a younger consultant named Aisha Rahman balanced a laptop on an upturned cable tray. They were investigating Birch. The graph showed supply air temperature, room temperature, occupancy, damper position, valve command, fan speed, carbon dioxide, outdoor air temperature, and a line labeled comfort score that had the serene cruelty of a number produced far from the room it described. "See?" Søren said. "Twenty-one point two degrees. Perfect." Aisha did not look convinced. "Where is the sensor?" "Above the display screen." "And where do people sit?" "Around the table, obviously." "Under the supply diffuser?" Søren leaned closer to the floor plan. "Mostly." Aisha folded her arms. "Then your room is comfortable at the wall and miserable at the humans." Søren grunted. He had spent thirty years learning that buildings lie most often through perfectly honest sensors. A thermostat in sunlight asks for cooling while people in shade reach for sweaters.

A carbon-dioxide sensor near a return grille reports average air while a crowded corner grows stale. A motion detector sees no one because people are sitting still. A temperature sensor says the zone is satisfied because the one square centimeter surrounding it is satisfied.

The ordinary thermostat is a small democracy with a very limited electorate. It measures one point and speaks for a room. In older buildings that point might command a boiler, a radiator valve, or a fan coil with almost comic simplicity: too cold, heat; too warm, stop heating; perhaps add cooling if the building is grand enough to afford summer indoors. Smart buildings do not escape this logic. They multiply it. Sensors feed controllers, controllers command actuators, actuators change flows of air, water, light, and motion, and supervisory software watches the watchers. A variable-air-volume box opens or closes to send more or less conditioned air into a zone. A chilled-water valve modulates to remove heat. A heat pump adjusts compressor speed. Blinds tilt to reduce solar gain. Lights dim when daylight is sufficient. Ventilation increases when carbon dioxide suggests people have arrived. The building is a stack of feedback loops laid over a stack of human activities. If the elevator chapter showed a building as a machine for moving bodies vertically, the smart-building chapter shows a building as a machine for editing the conditions under which bodies work, heal, learn, shop, sleep, and complain.

Nora complained at 9:04. By then twelve people had gathered in Birch for a meeting about winter respiratory illness. The irony was not lost on them when three participants kept their coats on. "Can we make it warmer?" asked Dr. Feld, visiting from the hospital. Malik, who had already loosened his collar, said, "Please do not." Nora walked to the wall panel, touched the up arrow, and watched a polite message appear: temporary adjustment accepted. "That means nothing," Malik said. "It means the building has heard us." "So has my teenage son." Ten minutes later the room still felt cold near the table and warm near the wall. The ventilation made a soft rushing sound above them. On the agenda, they were supposed to discuss vulnerable populations. In the room, vulnerability had become oddly literal. Dr. Feld said, "I cannot think about public health while my hands are freezing." Malik said, "I cannot think when it is stuffy." Nora looked at the ceiling. "The building is trying to satisfy all of us and has decided to satisfy none of us equally."

Comfort is not a number, though buildings must turn it into numbers to control it. Engineers have long used models of thermal comfort that combine air temperature, radiant temperature, humidity, air movement, clothing, and metabolic activity. These models are useful and necessary. They also reveal the problem. People are not identical instruments calibrated at the factory. A person who climbed stairs is not the same thermal object as a person who sat in a taxi. A nurse crossing a ward, a programmer staring at code, a child in a classroom, an elderly patient in a hospital bed, and a cleaner pushing a cart through a corridor all produce heat differently, dress differently, move differently, and tolerate discomfort differently. The building sees occupancy; it may not see fatigue. It sees temperature; it may not see dignity. It sees a scheduled meeting; it may not see that one person is afraid to ask for warmth because another person always jokes about sweaters. When comfort becomes averaged, the average body becomes a quiet political fiction.

This is why the politics of smart buildings begins before anything goes wrong. It begins with the set points. Who chooses twenty-one degrees rather than twenty-three? Who decides how much fresh air is enough when energy prices rise? Who has permission to open a window? Who can override the schedule? Whose phone app controls the lights, and whose presence is merely inferred? A landlord may want lower energy bills. A facilities team may want fewer complaints and fewer midnight alarms. A sustainability officer may want lower carbon emissions. A tenant may want prestige, certifications, and a lobby screen showing savings in green numbers. Workers may want air that does not make them sleepy, light that does not flatten their faces, and a room that does not require fingerless gloves. These desires are not always enemies, but they are not the same desire. A building management system converts them into priorities, deadbands, schedules, penalties, and constraints. It is a bureaucracy made of software and dampers.

At noon, Aisha joined Nora in Birch with a small handheld sensor and the expression of someone about to disappoint both humans and machines. Søren came with her. "We found the issue," he said. "The issue is that I am cold," Nora said. "That is one of the symptoms." Aisha placed the sensor on the table.

"The wall thermostat is near the screen, which warms during presentations. The controller thinks the room is slightly warmer than the occupied zone. Also, the air supply is washing down too directly on the table when occupancy is high." Malik pointed at the ceiling. "So the building is breathing on us wrong." "Yes," Aisha said. "That is the technical term." Nora laughed. Søren did not. He was thinking about ladder access, diffuser adjustment, sensor relocation, software permissions, and whether the contractor would claim the system had been delivered according to specification. "Can you fix it?" Dr. Feld asked. Aisha hesitated. "We can improve it. Buildings are rarely fixed. They are tuned, argued with, and maintained."

That sentence matters because smartness is often sold as installation, when it is really a relationship. A building is not smart on the day the sensors are commissioned and the dashboard goes live. It becomes smarter, or stupider, through use. Occupants move furniture under vents. Departments change schedules. A tenant adds computers and printers to a room designed for six people and then wonders why it overheats. Filters clog. Dampers stick. Software updates change interfaces. A security policy disables a data connection the energy optimizer needed. A pandemic changes the value of ventilation overnight. Electricity prices fluctuate. Weather grows less like the historical file used in design. The control problem is therefore not merely to regulate a stable object. It is to keep learning what the object has become. Søren understood this better than anyone in the building, though no brochure quoted him. "Every building is a prototype," he liked to say. "Some are just too expensive to admit it."

The promise, however, is real. Buildings consume a great deal of energy, and much of that energy is spent maintaining indoor conditions that are empty, excessive, poorly timed, or badly coordinated. A smart building can preheat before people arrive and coast when they leave. It can reduce ventilation in unused rooms while protecting air quality in crowded ones. It can dim lights near windows, lower blinds before overheating begins, use thermal mass to shift cooling away from expensive peak hours, detect faults before a pump fails, and coordinate with the electric grid so comfort is not purchased at the dirtiest moment. In a school, this may mean classrooms that are fresh when children arrive rather than overventilated all night. In a hospital, it may mean safer pressure relationships between rooms, alarms when humidity threatens equipment or infection control, and backup strategies when power fails. In an apartment building, it may mean heat delivered where people need it rather than wasted through overheated corridors and open windows. The moral case for smart buildings is that waste is not neutral. Energy wasted invisibly still becomes fuel burned somewhere, capacity built somewhere, emissions breathed somewhere, and money taken from someone.

Yet the same sensors that make waste visible can also make people visible. Occupancy sensing begins innocently: the room is empty, so turn off the lights; the floor is crowded, so increase ventilation; the elevator lobby is busy, so send more cars. But an occupied building is also a map of behavior. Badge data shows who arrived late. Wi-Fi associations suggest who lingered where. Desk sensors reveal which teams use space. Camera analytics count bodies. Meeting-room systems learn that some groups book rooms they do not use. In a workplace meeting about the civic center's new analytics package, a vendor named Clara presented heat maps in cheerful colors. "This helps optimize space utilization," she said. Nora looked at the glowing plan of her floor. "Can it show whether people are collaborating?" Clara brightened. "We can infer interaction patterns from co-presence." Henrik from security murmured, "That is a very expensive way to say spying." Clara's smile tightened. "The data is aggregated." Nora asked, "Aggregated for whom, and reversible by whom?" The room became colder without any change in temperature.

Control engineers know that observability is power. You cannot control what you cannot sense, but sensing changes the social meaning of the space being sensed. A library that counts visitors to adjust ventilation is different from a library that identifies which visitor sat beside which other visitor for how long. An office that learns occupancy to save energy is different from an office that uses the same data to punish remote work. A hospital that tracks equipment is different from a hospital that tracks nurses so aggressively that every pause looks like inefficiency. Smart buildings therefore force a question that thermostats never had to answer openly: how much should a building know about the people inside it? The answer cannot be left to procurement documents and privacy notices written in a language nobody reads. If comfort, energy, safety, and privacy are all control objectives, then privacy must be treated as a real constraint, not as a paragraph added after the sensors have already been installed.

By late afternoon, Birch had improved. Aisha had adjusted the airflow pattern temporarily, Søren had reduced the controller's eagerness, and the wall panel had been granted a slightly wider local override. The room was not perfect. Malik still found it warm; Nora no longer needed her coat. That was what success looked like in a shared environment: not paradise, but a truce. As people left, Nora stayed behind with Aisha and looked at the small sensor on the table. "Do you ever worry," she asked, "that we are making buildings too complicated for ordinary people to inhabit?" Aisha unplugged her laptop. "I worry more that we make them complicated and then pretend the complexity has disappeared." "What should people know?" "That the building is always making choices. That the defaults matter. That overrides matter. That maintenance matters. That a complaint is sometimes a sensor the system forgot to install." Nora considered this. "So my cold hands are data." "Yes," Aisha said. "But only if someone believes them."

At closing time, the civic center shifted into evening mode. Lights dimmed in offices, cleaning crews appeared, elevators changed their dispatch pattern, ventilation fell in empty zones, and the heating plant prepared for a colder night. The lobby screen announced that the building had saved energy compared with a conventional baseline. The number was probably true, within the assumptions that made it true. Henrik put on his coat at the security desk and watched Nora leave. "Birch fixed?" he asked. "Improved," she said. "That is what engineers call hope." Outside, the city was dark and windy. Inside, pumps slowed, valves moved by small degrees, dampers settled, sensors reported, algorithms predicted morning occupancy, and the building tried to dream of tomorrow's comfort in numbers. The lesson of smart buildings is not that buildings should decide what humans feel. It is that modern comfort is already a controlled condition, produced by machines, energy, measurements, and rules. When those rules become intelligent, they must also become accountable. A humane smart building would not be the one that never receives complaints. It would be the one that knows complaints are part of the loop, that comfort belongs to bodies rather than dashboards, and that efficiency is admirable only when it does not make people disappear into the average.

Chapter 15

Medicine by Feedback Loop

Wherever the art of medicine is loved, there is also a love of humanity.
—Hippocrates, *Precepts*, Section 6

In the winter light of Copenhagen, in the early 1920s, August Krogh stood beside his wife Marie and understood that science had become personal. He was already famous. In 1920 he had received the Nobel Prize in Physiology or Medicine for his work on the regulation of capillaries, the tiny vessels through which blood gives oxygen to tissue and carries waste away. He had spent years thinking about living control: how the body opens and closes pathways, how supply follows demand, how invisible mechanisms keep muscles and organs alive. But no prize protected a household from fear. Marie Krogh, herself a physician and scientist, had diabetes. Before insulin, that diagnosis was not a chronic inconvenience but a sentence written slowly in hunger, weakness, thirst, and decline. The treatment was often starvation diet, discipline without rescue, a desperate attempt to reduce sugar in the blood by reducing life itself. Imagine them at a table after the children have gone quiet, a medical letter lying open between them. "You know what this means," Marie says. August does not answer quickly. He has spent his life measuring gases, capillaries, exercise, metabolism. Now the variable that matters is not in a laboratory animal or a graph. It is in the woman across from him. "It means," he says at last, "that we must learn whether the Canadians are right."

The Canadians were Frederick Banting, Charles Best, J. J. R. Macleod, James Collip, and the group in Toronto who had isolated and prepared insulin in a form that could be used to treat people. News of their work crossed the Atlantic with the force of a miracle and the uncertainty of a rumor. Insulin was not a cure for diabetes, but it could turn imminent death into managed life. It replaced, from outside the body, a signal the body could no longer provide properly. August and Marie Krogh travelled to North America in 1922. In this chapter's imagined version, they sit in a Toronto office while snow darkens the windows and a young researcher brings in notes still smelling of ether and animal pancreas. "It is difficult to make," the researcher says. "Impure preparations can be dangerous. The dose matters. The timing matters. Food matters." Marie leans forward. "Everything in diabetes matters." August asks the question of a control engineer before the language has entered medicine in that form: "If the body no longer regulates sugar, how closely can we imitate the regulation?" The answer, in 1922, was crude but magnificent. Not closely enough to be natural. Closely enough to save lives.

When the Kroghs returned to Denmark, they did not merely bring hope home in a suitcase. They helped bring production. August Krogh obtained permission to manufacture insulin in Scandinavia, and with the physician H. C. Hagedorn and the pharmacist August Kongsted, he helped establish Nordisk Insulinlaboratorium in 1923. Marie's illness was one of the reasons the effort mattered so urgently; love and public health met in the same decision. The later history would branch, quarrel, innovate, and merge. The Pedersen brothers, Harald and Thorvald, who had worked with the early insulin enterprise, later founded Novo Terapeutisk Laboratorium after a dispute, and decades later Novo and Nordisk joined to form Novo Nordisk. But at the emotional beginning of the Danish insulin story stands a husband and wife facing a disease, and a Nobel laureate who understood that a discovery did not become medicine until someone could make it reliably, bottle it, distribute it, and teach patients how to use it. In an imagined Copenhagen laboratory, Hagedorn looks at the first difficult batches and says, "This is not like publishing a paper." Krogh replies, "No. A paper may be corrected next year. A vial is injected tomorrow."

That sentence contains the control problem of medicine. A body is already full of feedback loops. Blood glucose rises after a meal; beta cells in the pancreas secrete insulin; insulin helps tissues take up glucose and tells the liver to store rather than release it; glucose falls; insulin secretion falls. If glucose falls too far, other hormones push in the opposite direction. The healthy body is not a machine with one dial but a parliament of signals, delays, reserves, alarms, and compromises. Diabetes damages this regulation. In type 1 diabetes, the insulin-producing cells are destroyed, and insulin must be supplied from outside. In type 2 diabetes, the body may resist insulin's action and later fail to produce enough. Either way, the patient becomes part of the control loop. Food, exercise, infection, stress, sleep, menstruation, alcohol, fear, and growth can all change the system. The body is the plant, glucose is a measured variable, insulin is an actuator, meals are disturbances, and the patient is often the controller, estimator, alarm system, historian, and moral witness all at once.

Consider a Danish girl named Anna Sørensen in 1924, one of the first children in Copenhagen to receive insulin made by the new laboratory. Her father, a schoolteacher, watches the physician prepare the syringe with the frightened attention of a man seeing both a weapon and a sacrament. Anna is thin enough that her wrists look too large for her arms. Her mother holds a bowl of broth that Anna has not wanted. "Will it hurt?" Anna asks. "A little," the physician says. "Will I be well?" The physician looks at the parents before answering. "You may live. That is the first task." The injection is not automatic. It does not know what Anna ate. It does not know what she will eat tomorrow, whether she will run in the courtyard, whether fever will come in the night. The feedback loop is human and slow. Urine is tested. Symptoms are watched. Doses are adjusted by memory and caution. Too little insulin leaves sugar high and the body starving in the midst of plenty. Too much insulin drives sugar low and can bring sweating, trembling, confusion, seizure, coma. The treatment saves Anna by placing her family inside a new discipline of measurement and correction.

For much of the twentieth century, diabetes control was a conversation between imperfect measurements and bodily feeling. Patients boiled glass syringes, sharpened needles, measured urine sugar with tablets or strips, weighed food, counted portions, and learned the private weather of hypoglycemia. A man might say, "I know I am low because the room moves away from me." A child might say, "My hands feel full of bees." A nurse might say, "Never trust only courage. Test." The first great change was not automation but visibility. Blood glucose meters moved information from clinic to kitchen table. A patient no longer had to infer the present from yesterday's urine or from the trembling of his own hands. He could prick a finger, place a drop of blood on a strip, and see a number. That number was not wisdom, but it was feedback. It shortened the loop. It allowed correction before catastrophe. It also created a new burden: a life interrupted by measurement. In an imagined clinic in the 1980s, a teenager named Michael pushes a meter across the desk and says to Dr. Larsen, "Everyone says the number is for me, but it feels like a grade." Dr. Larsen answers, "Sometimes numbers become cruel because people forget they are instruments." "My mother cries when it is high." "Then we must teach your mother what the number means." "What does it mean?" "It means the next decision, not your worth."

Insulin itself also changed. Early extracts from animal pancreases kept people alive but varied in purity, duration, and predictability. Researchers learned to prolong insulin action, to purify preparations, to make human insulin through biotechnology, and later to design analogues that acted faster or slower. Each improvement was a change in the actuator. A controller is only as good as the thing it can command. If insulin acts too slowly, a meal can outrun it. If it lasts too long, yesterday's correction can become tonight's emergency. Long-acting insulin tried to provide a background level, like a quiet basal signal. Rapid-acting insulin tried to meet meals, like a quick control action against a disturbance. Patients and clinicians began to speak, sometimes without knowing the engineering word, in terms of profiles, peaks, delays, sensitivity, correction factors, and targets. The problem was not simply to lower glucose. It was to lower it at the right speed, at the right time, without causing a dangerous undershoot. Medicine had become a problem of dynamic control.

The insulin pump made the loop more mechanical. Instead of one or several injections each day, a small device could deliver insulin through a tube or patch in tiny programmed amounts. A engineer named Sara Holm, whose younger brother has diabetes, first sees a pump clipped to a patient's belt and says, "So it is a pancreas?" The diabetes nurse laughs softly. "No. It is a very obedient servant with no judgment." The pump can deliver basal insulin hour by hour and boluses for meals, but someone must tell it what the meal means. How many carbohydrates? How much insulin per gram? Is there active insulin still working from the last correction? Will the patient walk home in the rain and burn glucose faster? The pump improves control because it makes insulin delivery more flexible, but it also creates new failure modes: blocked infusion sets, empty reservoirs, wrong settings, buttons pressed in fatigue, software misunderstood, alarms ignored under blankets at two in the morning. A syringe is simple and limited. A pump is powerful and needy. It asks the patient to trust a machine attached to the skin.

Then came continuous glucose monitoring, and the measured variable began to speak almost in real time. A sensor under the skin estimates glucose in interstitial fluid and sends readings every few minutes. The patient sees not only a number but an arrow: rising, falling, steady, plunging. This is a profound change in the experience of disease. A finger-stick meter is a photograph. A continuous monitor is weather radar. It reveals the night, the hidden rise after breakfast, the delayed fall after exercise, the way pizza behaves like a slow conspiracy. In an imagined bedroom, a mother named Lina wakes to an alarm from her son's monitor. The phone says he is low. Her husband whispers, "Again?" She is already standing. In the boy's room, twelve-year-old Emil is half awake and irritated. "I hate the beeping." "I know," Lina says, holding out juice. "It is trying to help." "It is spying." "Tonight spying is help." He drinks, angry and alive. The alarm is feedback with a voice, and like every voice it can become exhausting. Too many alarms and the family sleeps through them. Too few and danger passes silently. The design of alarm thresholds is not merely technical. It decides how fear enters a household.

Closed-loop insulin systems, often called artificial pancreas systems, joined pump and sensor with an algorithm. The sensor estimates glucose; the controller predicts where glucose is going; the pump changes insulin delivery. Some systems suspend insulin when glucose is predicted to go low. Others increase or decrease basal delivery and suggest or automate corrections. The dream is obvious: let the machine do more of what the pancreas once did. Let the patient sleep. Let parents stop standing guard over childhood. Let numbers become background again. But the phrase artificial pancreas can mislead. The real pancreas senses blood directly, secretes insulin into the portal circulation, coordinates with other hormones, and responds through biology refined by evolution. A pump under the skin delivers insulin into tissue, where absorption is delayed and variable. The sensor is delayed too. The meal is often announced late or estimated badly. The controller is always chasing a body that changes its rules. In control terms, the system has measurement delay, actuation delay, unmeasured disturbances, nonlinear response, saturation, safety constraints, and a human who may or may not cooperate. In human terms, it is Tuesday night, the child ate cake, football practice was harder than expected, the sensor adhesive is peeling, and everyone is tired.

That is why closed-loop medicine is both beautiful and humbling. Imagine Sara Holm again, now working for a medical device company, testing an algorithm with Dr. Larsen in a simulated clinic. On the screen, a glucose trace rises after an unannounced meal. The controller increases insulin, then hesitates because too much insulin too late could cause a crash. "It looks indecisive," Sara says. Dr. Larsen folds his arms. "Good." "Good?" "A decisive fool is dangerous." Sara points to the curve. "If we wait, the glucose stays high." "If we chase, the patient may be low at midnight." The engineer hears in that sentence the old control lesson: correction is not virtue by itself. The timing, magnitude, and uncertainty of correction matter. Medicine adds a further demand. A high glucose value is not just an error signal; a low glucose value is not just an overshoot. These are sensations, risks, hospital visits, parents' marriages, school trips, sleep, shame, independence, and cost. The controller touches a life.

Feedback medicine reaches beyond diabetes. A pacemaker listens for the heart's rhythm and intervenes when the heart is too slow or dangerously disordered. Implantable defibrillators wait in terrifying patience, ready to deliver a shock when rhythm becomes lethal. In anesthesia, clinicians watch blood pressure, oxygen saturation, expired gases, brain activity, and movement, adjusting drugs so that a patient is neither awake in pain nor driven too deep into danger. In intensive care, ventilators, infusion pumps, dialysis machines, oxygen controllers, and alarms create a forest of partial loops around a fragile person. Each loop has a target. Each target belongs to a body that may not care about our neat separation of variables. Raise pressure and you may stress the heart. Give fluid and you may flood the lungs. Increase oxygen and you may cause injury. Suppress pain and you may suppress breathing. In an operating room, an anesthesiologist named Farah watches a monitor while a young resident says, "The machine can maintain the concentration automatically." Farah nods. "It can maintain a concentration. I maintain a patient."

The moral question in medicine by feedback loop is not whether automation should exist. It already does, and much of it is merciful. The question is how much authority to give it, how transparently it should act, how it should fail, and who remains responsible when the loop closes around a body. A device may know the target range but not the family argument that made a teenager skip lunch. It may detect a rhythm but not the patient's wish to avoid shocks near the end of life. It may maintain blood pressure but not understand that surgery has changed suddenly from routine to catastrophic. The human in the loop is not always wiser, calmer, or faster than the machine. Humans forget, misread, tire, and deny. But humans can sometimes understand meaning where sensors see only signal. Good medical control therefore does not simply replace people. It reallocates vigilance. It lets machines do what machines do well: measure continuously, remember settings, calculate trends, respond at night, avoid boredom. It asks people to do what only people can do well: interpret goals, notice context, consent, comfort, doubt, and decide when a target is no longer the point.

Near the end of her life, imagine Marie Krogh looking at a child in Copenhagen who has survived because insulin exists in a vial. The girl is not cured. She will have to measure, inject, eat carefully, listen to her body, and negotiate with a disease that never sleeps. But she will grow. She may become a teacher, a pianist, a mother, an engineer, a difficult old woman with opinions. August stands beside Marie and watches the child leave the clinic with her father. "It is not enough," Marie says. "No," August answers. "But it is a beginning." A century later, a child like Emil sleeps under the guard of a sensor, an algorithm, and a pump, while his mother sleeps lightly in the next room and a phone waits to alarm. The loop is faster now, more automated, more mathematical, more commercial, more expensive, more capable. Yet the old reason remains visible. The purpose of feedback in medicine is not to make elegant graphs. It is to give time back to the person whose body has lost one of its own ways of keeping time. It began, in the Danish insulin story, with a scientist trying to help his wife live. It continues whenever a machine listens to a body and answers carefully enough that a human being can return, for a while, to the ordinary business of being alive.

Chapter 16

Robots Among Us

The machine does not isolate man from the great problems of nature but plunges him more deeply into them.

—Antoine de Saint-Exupéry, *Wind, Sand and Stars*, Chapter III

Before most of Gothenburg had finished dreaming, a robot arm in a car factory outside the city came alive in the narrow technical sense that machines do. It did not wake in any human sense. No thought rose inside it. No reluctance, ambition, hunger, or memory crossed its controller. A signal arrived, drives enabled, brakes released, and the arm moved from its rest position with the polished certainty of a machine that had repeated the same gesture so often that repetition had become its form of existence. Beside the safety fence, a technician named Lars Nyström watched with a mug of coffee in one hand and a teach pendant in the other. A new apprentice, Emil, stood beside him wearing ear protection slightly too large for his head. The arm lifted a panel, turned it, presented it to a fixture, and waited while another machine clamped it. "Does it ever get bored?" Emil asked. Lars looked at him. "That is why it is employed." "Doing the same thing all day?" "Doing the same thing exactly all day. Humans are not very good at that. We either suffer or improvise." The robot lowered its wrist, followed a seam, and the welding arc flashed blue-white behind tinted glass. "And robots?" Emil asked. Lars took a sip of coffee. "Robots suffer only when engineers are lazy."

Industrial robotics is where many people first met the practical meaning of a robot: not a metal person, but a programmable machine able to move tools and parts through space with repeatable accuracy. The first great workplace of robotics was not the living room or the hospital corridor but the factory, where dullness, danger, speed, heat, fumes, mass, and precision made mechanical labor valuable. A robot could weld a seam without blinking at the arc, paint a body shell without breathing solvent, lift a heavy casting without ruining its back, place components with a steadiness no tired hand could maintain, and repeat a motion thousands of times while producing variation small enough for modern manufacturing to trust. Companies such as ABB Robotics helped make this world ordinary: orange arms in cells, controllers in cabinets, teach pendants in operators' hands, software planning paths, and factories reorganized around machines that could work with a discipline neither animal nor human. The contribution was not merely replacing muscle. It was making accuracy programmable, making quality repeatable, and making hazardous work less dependent on bodies that could be burned, crushed, poisoned, or exhausted.

A robot arm is a control problem made visible. Each joint must know where it is, where it should go, how fast it may move, how much torque is needed, what load is attached, whether a command would exceed limits, and how to arrive without shaking the tool or striking the world. Encoders measure position. Motors apply torque. Controllers compare desired and actual motion many times per second. Models compensate for gravity, inertia, friction, backlash, compliance, and payload. Safety systems monitor doors, light curtains, speed limits, emergency stops, and sometimes the presence of people nearby. To an observer, the robot seems graceful because the ugly arithmetic has been hidden inside cabinets and code. Emil learned this the first time Lars let him jog an arm slowly through a new path. "It feels alive," Emil said. Lars shook his head. "It feels controlled. Do not confuse those. Alive forgives you sometimes. Controlled obeys exactly enough to hurt you if you told it the wrong thing." That is the factory lesson. Precision is a gift only when surrounded by limits.

The economic promise of industrial robots has always been double-edged. They increase efficiency by working quickly, consistently, and without the rhythms of fatigue. They improve accuracy by reducing variation. They can lower costs, raise output, reduce scrap, and allow products to be made at tolerances that would be unreasonable if every motion depended on human steadiness. They can also displace workers, change skill requirements, concentrate value in firms that can afford automation, and make production feel less like craft and more like supervision of invisible decisions. In the lunchroom, Emil says, "My uncle says robots take jobs." Lars unwraps a sandwich. "Some jobs, yes." "Then why are we proud of them?" "Because some jobs take people." Emil waits. Lars points with his sandwich toward the welding hall. "That place used to eat lungs and shoulders. The question is not whether robots should do work. The question is what work humans should be asked to do, and what society owes the people whose work changes." A robot can save a worker from danger and still leave a family afraid of the future. Both truths can stand in the same factory.

Robotics did not remain bolted to the floor. Once engineers asked machines to move through the world rather than only move within a fenced cell, the control problem became harder and more interesting. A walking robot must balance, step, recover, sense terrain, estimate its body state, and decide where a foot can safely land. Wheels are efficient but limited by stairs, rubble, mud, forest, and human architecture. Legs are complicated but promise access to places built for animals and people. In an imagined test field, a robotics researcher named Sofia Berg watches a quadruped robot pick its way over wet stones while a rescue worker named Henrik follows with folded arms. The robot slips, catches itself, and continues. Henrik says, "It walks like a nervous dog." Sofia smiles. "A nervous dog with lidar, inertial sensors, joint torque control, and no sense of pride." "Can it enter a collapsed building?" "That is the hope." "Hope is not a rescue plan." "No. But neither is sending a person into a void we have not mapped." The machine takes another step, places a foot badly, corrects, and stops. Henrik nods despite himself. "If it falls through the floor, I will feel bad for the budget. Not for its children."

That is one of the noblest uses of robotics: sending machines where the environment is hostile to human bodies. Robots can inspect nuclear facilities, mines, sewers, offshore structures, burning buildings, chemical spills, underwater pipelines, wind turbines, and disaster zones. They can defuse bombs, map radiation, crawl through pipes, carry cameras into unstable ruins, and lift loads in warehouses until a human spine no longer has to become the cheapest actuator. They can work in sterile spaces, clean rooms, slaughterhouses, foundries, and laboratories where contamination, heat, repetition, or danger makes the human presence costly in more than money. The moral argument is strongest when the robot takes over tasks that are not meaningful human work but organized injury: breathe this dust, lift this weight, enter this smoke, hold this tool against vibration for thirty years, stand close to this blast because someone must. Control then becomes protection by distance. The operator remains in the loop, but the fragile body is moved away from the hazard.

Humanoid robots enter the story with a different promise and a greater temptation. A robot shaped roughly like a person can, in principle, use spaces, tools, stairs, doors, handles, and furniture designed for people. It may carry boxes in a warehouse, fetch objects in a home, guide visitors in a hospital, assist in a laboratory, or work in a shop after closing. But the human shape carries emotional weight. People expect faces to understand, hands to intend, voices to care, and eyes to notice. A humanoid that can walk, gesture, answer questions, and carry a tray may invite trust faster than its competence deserves. In a demonstration hall, a product manager named Clara asks a humanoid robot to bring a bottle of water to a table. It walks slowly, stops, turns, extends an arm, and places the bottle down with ceremonial caution. An investor applauds. A safety engineer named Nils whispers to Clara, "It took three minutes to do what a bored intern does in eight seconds." Clara whispers back, "The intern cannot raise capital." Nils watches the robot reset its posture. "Then let us make sure the capital does not forget physics." Humanoids are impressive not because they are almost human, but because being even slightly useful in a human world is brutally difficult.

The technical problems multiply near people. A robot in a cage can move fast because people are excluded. A robot in a home, hospital, school, or shop must treat uncertainty as normal. Children run unpredictably. Pets sleep in paths. Floors are cluttered. A cup may be full, empty, hot, fragile, or precious. A hand offered for support belongs to a person who can be hurt. Service robots therefore require perception, planning, force control, compliant actuation, speech interfaces, privacy safeguards, and a humility that industrial machines never needed. The robot must not merely reach a target. It must ask whether reaching is appropriate. It must know when to slow down, when to stop, when to call for help, and when a human instruction should be refused. "Pick up that knife," a person may say. In a kitchen this may be ordinary. In a quarrel it may not be. The domestic robot inherits the oldest problem in control: the set point is not enough. Context decides whether the command is wise.

Personal assistant robots make this question intimate. A useful assistant might carry laundry, remind someone to take medicine, fetch dropped objects, open doors, bring food, guide a person through exercises, read labels, connect a video call, monitor whether the stove is on, or summon help after a fall. For a busy parent, a disabled adult, a hospital nurse, or a person living alone, such help could mean less fatigue and more independence. In an imagined apartment in Malmö, an elderly widower named Arne tests a home assistant robot provided through a municipal care program. His daughter, Liv, watches suspiciously while the robot rolls to the kitchen. "Arne, it is time for your afternoon tablets," the robot says. Arne looks at Liv. "It is more polite than you." Liv says, "I am not programmed to flatter you." The robot opens a compartment. Arne takes the pills and water. "Can it make coffee?" he asks. "Not yet," Liv says. "Then civilization remains incomplete." He smiles, but Liv does not. "I do not want them replacing visits with this thing." Arne's face changes. "I do not want you driving across town because I dropped a spoon."

There is the dilemma. Assistance can be freedom. A robot that helps an elderly person rise from a chair, carry groceries, remember medication, or call for help can reduce dependence on hurried relatives or overworked care staff. It can allow a person to remain at home longer, preserve routines, and ask for small help without embarrassment. Many people would rather ask a machine ten times to repeat an instruction than feel they are burdening a daughter, son, nurse, or neighbor. Dependence on humans can be loving, but it can also feel humiliating when every minor need becomes a negotiation with someone else's time. Arne tells Liv this one evening after the robot has retrieved his glasses from under the sofa. "You think I am lonely because the machine helps me," he says. "Sometimes I am lonely because I need help before I can decide anything else." Liv sits beside him. "I want you to need people." "I do need people. I do not need people for every glass of water." That distinction is easy to miss from outside old age.

But assistance can also become abandonment with better engineering. If a municipality, family, insurance company, or hospital treats a robot as a substitute for human presence rather than a tool that supports it, the machine may quietly widen loneliness. A robot can remind, lift, fetch, and monitor, but it cannot replace the moral fact of being visited by someone who could have chosen otherwise. It can simulate patience without caring, answer questions without understanding, and say good morning without noticing whether the morning is good. Some people may form attachments to robot companions, and those attachments may comfort them. The ethical question is not whether such comfort is false in a simple sense. Human beings have always been comforted by pets, photographs, letters, rituals, radio voices, and imagined presences. The harder question is who benefits when mechanical companionship is offered. If it gives Arne more control over his day and frees Liv's visits from chores so they can talk, it may be humane. If it lets the care system reduce visits until the robot becomes the only regular voice in the room, efficiency has disguised neglect as innovation.

Robotics therefore brings together many themes of control. It is feedback in motion: sensors, actuators, models, estimation, planning, correction, and safety. It is also society in motion: labor, dignity, danger, care, cost, privacy, and trust. Industrial robots show that control can make production faster and more accurate. Walking robots show that control can carry machines into terrain too uncertain for wheels and too dangerous for people. Humanoids show both the ambition and the theatrical risk of building machines in our own image. Personal assistants show that the most difficult environment is not the factory floor or the rubble pile but the human household, where every object has a history and every request has a social meaning. At the end of the Malmö trial, the robot reminds Arne of a video call. Liv appears on the screen from her kitchen, tired after work. "Did it behave?" she asks. Arne looks at the machine, then at his daughter. "It helped." "And are you lonely?" He considers lying because parents protect children even when children are grown. "Sometimes," he says. "But not because of the robot." Liv nods slowly. "Then I will come tomorrow." Arne smiles. "Come for coffee, not because I dropped something." The robot waits beside them, silent and useful. That may be the proper place for many robots: not replacing the human loop, not pretending to be the human loop, but reducing the needless burdens inside it so that people have more room for the parts of care that only people can give.

Chapter 17

The Algorithm That Controls Your Feed

My experience is what I agree to attend to.

—William James, *The Principles of Psychology*, Volume I, Chapter XI

Under a blue blanket in Aarhus on a school night, a sixteen-year-old named Maya Jensen lies with one thumb moving through a feed that seems to know her better than her friends do. The room is dark except for the phone. Her mathematics book is open on the desk, face down like a small defeated animal. Every few seconds the screen offers a new fragment of the world: a dog failing nobly to climb onto a sofa, a comedian mocking teachers, a girl with perfect skin explaining morning routines, a war clip cut to dramatic music, a recipe, a political joke, a singer crying in a bathroom, a boy lifting weights, a stranger saying that if you feel empty there is a reason. Maya does not choose the sequence in any ordinary sense. She does not turn pages, tune a radio, or select a channel. She performs small gestures, and the system answers. Pause. Swipe. Rewatch. Like. Skip. Follow. Search. Share. Hover. Return. The phone does not ask,

"What do you want to become?" It asks a narrower question millions of times: "What will keep you here for the next few seconds?" Downstairs, her father calls, "Maya, sleep." She says, "In a minute." The algorithm has heard that sentence from humanity many times and has built an empire out of the minute.

Inside the company that runs the feed, let us imagine an engineer named Daniel Park sitting in a glass meeting room in California while a product manager named Elena Ruiz points at a dashboard. Neither is a villain. Daniel has spent years building systems that help people find things they like. Elena believes, sincerely, that a good feed can make the world less lonely. On the screen are curves: watch time, daily active users, session length, retention, creator satisfaction, reports, removals, revenue. "The new ranking model improves seven-day retention," Elena says. Daniel looks at another column. "It also increases borderline political content in the test group." "Borderline by whose definition?" "By our classifier's definition, which is not a comforting answer." Elena rubs her forehead. "If we down-rank too hard, people say we are censoring them. If we do not down-rank, journalists say we are radicalizing them. If we show only safe material, users leave." Daniel says, "Then maybe the objective function is wrong." Elena gives him the tired look of someone who has said the same sentence in other rooms. "The objective function is never one thing. It is the argument we are having, written in numbers."

A recommendation feed is not a thermostat, and it is important not to pretend that it is. A thermostat measures temperature, compares it with a set point, and turns heating or cooling on and off. A social media feed measures traces of human behavior and uses them to rank possible posts, videos, advertisements, comments, and notifications. It does not control attention with gears or valves. It controls by selection. Out of an enormous reservoir of possible content, it chooses the next thing most likely to produce some desired response. The measured variable may be watch time, click-through rate, likes, comments, shares, replies, purchases, follows, or the probability that a user will return tomorrow. The actuator is not a motor but a ranking decision: put this post first, hide that one, send a notification now, recommend this channel, autoplay the next video, show the angry comment near the top because anger predicts reply. The disturbance is everything human life brings: boredom, grief, lust, politics, insomnia, adolescence, loneliness, curiosity, prejudice, humor, and hunger for belonging. The feedback loop is simple to describe and difficult to survive. The user acts; the system learns; the system changes what the user sees; the user acts again.

There is a good version of this story, and it should be told honestly. Before feeds became personalized, the internet was also full of frustration. Search results missed the thing you wanted. Forums were hard to find. Artists posted into emptiness. A teenager with a rare illness might never meet another person who understood. A musician in Lagos, a carpenter in Krakow, a mathematics teacher in Mumbai, and a grandmother in Odense could create something valuable and be discovered without owning a newspaper, a television station, or a publishing house. Recommendation systems can reduce the tyranny of the already famous. They can translate abundance into relevance. They can help a student find a lecture, a patient find a community, a cook find a technique, a citizen find emergency information, a lonely person find evidence that loneliness is not unique. In an imagined apartment in Manchester, an elderly widower named George watches videos of people restoring old radios because the algorithm noticed he lingered on one repair. His daughter teases him. "So now the machine knows you like dusty wires." George smiles. "Better than my doctor does." "That is not reassuring." "No, but it found me a man in Canada fixing the same Bush receiver I had in 1962." For George, the feed is not a monster. It is a companion librarian with strange instincts.

The same mechanism can also be generous to creators. A young woman with a sewing machine can find customers. A science communicator can reach millions without a broadcaster. A local protest can become visible. During disasters, feeds can surface warnings faster than official channels, though they can also surface rumors just as fast. Small businesses depend on recommendation and advertising systems to find people who might actually want what they sell. For many users, personalization is not manipulation but relief. They do not want the whole internet. They want the part that matters to them. Daniel says this in the meeting when the discussion turns dark. "We keep talking as if ranking is only harm," he says. "Without ranking, the feed is noise." Elena nods. "Exactly. The question is not whether to rank. The question is what our ranking makes more likely in the world."

That question became harder to avoid as platforms grew large enough for their internal experiments to become social facts. In 2014, researchers from Facebook and Cornell published a study in the Proceedings of the National Academy of Sciences describing an experiment in which the emotional content of some users' News Feeds was altered to study whether emotion could spread through the network. The effect reported was small, but the reaction was large because the study made visible something users had not fully absorbed: the feed was not merely showing life; it was arranging life, and that arrangement could be used to test mood. Imagine a Facebook user named Claire reading about the study over breakfast. "They changed people's feeds to see if they could make them sadder?" she asks her husband. "Sadder by a tiny amount, apparently." "That is not the defense you think it is." "The terms of service probably allowed it." Claire puts down the newspaper. "I am not asking whether a lawyer allowed it. I am asking when my feelings became a laboratory." The scandal was not that one experiment had transformed society. It was that the experiment revealed the ordinary power of ranking. If changing the mixture of positive and negative posts can measurably change expression, then the everyday feed is already an emotional environment designed by someone, or by something, in pursuit of a goal.

YouTube made the issue vivid in another way. Its recommendation system has helped people learn languages, repair bicycles, study calculus, discover music, and watch lectures that no television channel would have carried. It has also been accused, by journalists, researchers, and former insiders, of leading some users toward more extreme material because intense content can hold attention. The simple phrase rabbit hole became a public diagnosis of a control loop: watch one video, receive a stronger version, watch that, receive something more charged, until the system seems to have mistaken escalation for relevance. The company has said many times that it changed its systems to reduce recommendations of harmful borderline content, and these changes matter. Algorithms are not fixed demons. They are designed, measured, revised, and contested. But the YouTube debate showed that a recommender can optimize locally while changing a person globally. It may not intend to radicalize anyone. It may only learn that outrage, certainty, conspiracy, humiliation, and tribal identity are sticky. In control language, the system has found a high-gain path through human attention. In human language, a bored man looking for news about immigration at midnight may wake six months later convinced that enemies are everywhere.

TikTok made the loop faster and more intimate. The For You page reduces choice almost to a pulse. One video fills the screen; one gesture rejects it; another accepts it; the next arrives instantly. The Wall Street Journal's 2021 investigation used automated accounts to show how quickly the system could infer interests from watch time and feed accounts narrow streams of related content, including content about depression or sexual themes. TikTok has disputed aspects of outside characterizations and has added controls and safety features, but the central engineering lesson remains clear: when feedback is fast, adaptation is fast. The system does not need a user to declare, "I am anxious about my body." It can notice that she stops scrolling on body-checking videos. It does not need a boy to say, "I feel humiliated and angry." It can notice that he replays clips about dominance, revenge, and contempt. Maya, under her blanket, does not think she is training a model. She thinks she is passing time. But passing time is data, and data is the language in which the feed is corrected.

The bad in these systems therefore does not come only from bad content. It comes from a mismatch between what can be measured and what matters. Engagement is measurable. A person's slow change in self-respect is harder to measure. Watch time is measurable. Civic trust is harder. Clicks are measurable. Whether a teenager sleeps, whether a lonely adult becomes more isolated, whether a democracy becomes more suspicious of shared facts, whether a creator burns out trying to satisfy an invisible ranking machine: these are not absent because companies do not care at all. They are absent because they are difficult to turn into immediate feedback signals. A control system tends to improve what it can sense and score. If the sensor is narrow, the world is narrowed to fit it. Elena says this in another meeting, later, after a public controversy. "We optimized for meaningful social interaction," she says. Daniel answers, "Meaningful according to replies." "Replies are signal." "So are fistfights." Nobody laughs, because everyone in the room knows he is not joking.

The Facebook Papers and the testimony of whistleblower Frances Haugen in 2021 pushed this conflict into public view. Internal documents reported by journalists suggested that Meta knew significant things about the harms of its platforms, including problems connected with Instagram and teenage well-being, while also facing business incentives to keep people engaged. Meta disputed some interpretations and argued that the public reporting lacked context, but the broader issue became unavoidable: platforms are not neutral pipes when their ranking systems decide what billions of people see. A pipe carries what enters it. A feed chooses, orders, amplifies, suppresses, and repeats. In a Senate hearing, imagine a mother named Rachel Kim listening as executives speak about safety investments and user choice. Her daughter has not been destroyed by an app, not in a dramatic movie sense, but she has become harder to reach. Dinner is quieter. Sleep is shorter. The mirror is more powerful. Rachel whispers to another parent, "They keep saying the users choose." The other parent says, "My daughter chooses too. But the machine chooses what she chooses from." That is the deep control problem. Freedom inside a designed environment is still shaped by the environment.

There are also political cases where feeds made reality feel unstable. During elections, pandemics, wars, and protests, recommendation systems have had to decide what to do with misinformation, state propaganda, satire, ordinary error, and legitimate dissent. Remove too much, and platforms become private ministries of truth. Remove too little, and lies scale faster than corrections. Rank official sources higher, and some users accuse the system of protecting power. Rank popularity higher, and dramatic falsehoods may outrun boring accuracy. During the COVID-19 pandemic, platforms promoted public health information while also fighting waves of misinformation about vaccines, masks, cures, and conspiracies. The feed became a medical environment without being a hospital, a political environment without being a parliament, an educational environment without being a school. It had public consequences but private governance. In an imagined policy call, a trust-and-safety lead named Priya says, "If we demote this claim, we will be accused of censorship." Daniel says, "If we promote it, people may drink bleach." Elena says, "The model does not know bleach." Priya answers, "Then we had better not pretend the model is governing alone."

The analogy with control engineering helps because it gives us a disciplined way to ask what is being controlled. In classical control, a careless objective can destroy a system. Tell an autopilot only to minimize altitude error and ignore passenger comfort, actuator limits, and stall margins, and it may command dangerous motions. Tell a recommender only to maximize engagement and it may discover that fear, lust, anger, envy, and belonging are powerful actuators. Engineers know that every controller needs constraints, robustness, monitoring, and failure analysis. The feed needs the same. What content is outside the admissible set no matter how engaging it is? What kinds of repeated exposure should trigger diversity rather than narrowing? When should the system slow down instead of speed up? How should it distinguish a user's immediate impulse from a user's longer-term welfare? What rights should users have to inspect, reset, or alter the controller? What should be optimized for children, whose preferences are real but still forming? A thermostat has a set point. A society must argue about its set points in public.

The good systems of the future may not be the ones that know us most intimately, but the ones that allow us to negotiate with their knowledge. Imagine Maya five years later, now a university student, opening a feed that begins not with endless autoplay but with a small set of controls. "What do you want from this session?" it asks. Relaxation. Friends. News. Learning. Surprise. Stop me after twenty minutes. Avoid body-image content tonight. Show me why this was recommended. Let me choose chronological order. Let me see something outside my habits. Such controls would not solve every problem, and many users might ignore them. But they would change the moral shape of the loop. They would make the user less like a measured disturbance and more like a participant in control. Daniel argues for something like this in the company cafeteria. Elena says, "Friction reduces engagement." Daniel says, "So do seatbelts, if your metric is how fast people can leave the driveway." "That metaphor will annoy leadership." "Good. It annoys me too."

Regulation has begun to enter this space because private optimization has public effects. The European Union's Digital Services Act, for example, requires large platforms to assess and mitigate systemic risks and provide more transparency around recommender systems. Other jurisdictions debate child safety, data protection, algorithmic accountability, and competition. Regulation will not write the perfect feed, and governments can misuse control as readily as companies can. But the absence of regulation is also a choice: it leaves the social set points to advertising markets, growth targets, and internal dashboards. The point is not that every feed should become dull, official, and harmless. Human beings need play, strangeness, argument, desire, comedy, and discovery. A perfectly safe feed might be a dead feed, and a dead feed would not serve culture. The point is that systems with the power to steer attention should be designed with an account of human flourishing larger than the next click.

At 12:36, Maya finally turns the phone face down. The room does not become silent immediately, because the feed continues inside her for a while. A song repeats. A joke returns. A girl's face compares itself with hers in the dark glass of the window. Downstairs, her father has fallen asleep on the sofa waiting for the house to settle. In California, Daniel's experiment has finished, and the dashboard says the new model is a success by three measures and a worry by two. In Brussels, a regulator writes language about transparency. In a studio apartment, a musician wakes to find that the algorithm has carried her song across the world. In another room, a man is being shown the next proof that everyone unlike him is an enemy. The feed is not one story. It is a control system made of many stories, each small enough to look like choice and large enough, in aggregate, to reshape culture. The lesson is not to flee all recommendation. We cannot live in abundance without filters. The lesson is to ask who designs the filter, what it senses, what it rewards, what it hides, how it fails, and whether the person inside the loop is being helped to live or merely being kept from leaving.

Chapter 18

The Human in the Loop

The real problem is not whether machines think but whether men do.

—B. F. Skinner, *Contingencies of Reinforcement*, Chapter 9

The control room was quiet in the way important rooms become quiet after people have learned that panic wastes oxygen. Screens covered one wall. Trend lines moved with the patient confidence of machines that did not know the meeting was about them. A senior operator named Ingrid Solheim sat with her hands folded while a younger engineer, Daniel Price, explained a proposed automation upgrade to a committee of managers, safety specialists, and exhausted people who had spent their careers being awakened by alarms. The plant could have been a power station, a refinery, a water-treatment facility, or a large hospital energy center; let it stand for all of them. Daniel pointed to a display. "The new supervisory controller will detect abnormal patterns earlier than the operators can. It will recommend corrective actions, and in some cases it can execute them automatically." A manager nodded with the expression of someone hearing savings. Ingrid did not nod. "Earlier than which operators?" she asked. Daniel paused. "Earlier than a human monitoring dozens of variables." "That was not my question." The room looked at her. "Earlier than a tired trainee at three in the morning? Probably. Earlier than someone who has smelled this plant for twenty-five years? Perhaps not."

This chapter has been present from the beginning, even before it received a title. Thales carried the water clock, but Philon still had to decide when fairness had been served. Brother Martin listened to the tower clock, but Jean still climbed into the mechanism when rhythm became suspicion. Tom Price trusted the governor enough to let it help and distrusted it enough to keep it honest. Rick and Denise watched a grid whose alarms failed to tell them what the physics was already doing. The Chernobyl operators sat inside a control room where procedure, hierarchy, fatigue, and design flaw met in the worst possible way. The Gripen pilot moved in a sky where computers were essential and yet the relation between pilot command and flight-control law became fatal. The self-driving car chapter asked what happens when a machine must decide whether stopping is caution or obstruction. The feed chapter asked what happens when the human in the loop is not an operator but a user being operated upon. Again and again, the question returns: when should the person command, when should the person supervise, when should the person be protected from commanding, and when should the person refuse the loop entirely?

The phrase human in the loop can sound comforting because it seems to promise responsibility. If a person remains somewhere in the system, then surely judgment remains. But the phrase hides several different arrangements. A human can be in direct manual control, as Tom once was with the steam valve. A human can supervise automation, as a grid operator watches frequency and power flows. A human can approve recommendations, as a doctor may review an insulin dose, a radiology result, or an alarm. A human can intervene only after automation has failed, as a driver is asked to take over from a system that has been handling the road for minutes or hours. A human can be used as legal decoration: present enough to blame, not informed enough to decide. In the meeting, Daniel says, "There will always be a human override." Ingrid looks at the diagram. "Where?" "Here." "After the controller has already closed two valves, changed the pump schedule, and suppressed duplicate alarms?" "Yes." "Then the human is not in the loop. The human is standing beside the loop with a broom, waiting to sweep up." Nobody laughs, because the sentence is too useful.

Norbert Wiener gave the modern world a language for this problem when he helped name and shape cybernetics: the study of control and communication in animals and machines. The word mattered because it refused a simple boundary. A thermostat, an anti-aircraft predictor, a nervous system, a company, a government, a pilot, and a machine tool could all be understood through feedback, information, correction, and purpose. Wiener saw that machines and humans were not separate kingdoms. They could be coupled into systems, and the coupling could amplify wisdom or stupidity. He also warned, in his way, that a machine pursuing a goal literally may produce consequences no reasonable person intended. In an imagined seminar late in his life, let us place Wiener beside a young engineer named Ruth, who has built a controller that performs beautifully in simulation. "It minimizes the error," Ruth says. Wiener asks, "Whose error?" "The plant output." "And if the plant output is not the whole problem?" "Then we add constraints." "And if the constraints omit what the human beings meant but failed to say?" Ruth frowns. "Then the machine does what we asked, not what we wanted." Wiener nods. "That distinction is where ethics enters engineering."

Artificial intelligence has made this distinction larger and less avoidable. Many modern AI systems are trained on data: images, text, driving scenes, medical records, sensor histories, maintenance logs, user behavior, weather patterns, conversations, and countless examples of what happened before. They learn statistical regularities. They predict likely labels, likely next words, likely trajectories, likely failures, likely clicks, likely diagnoses, likely actions. That ability is powerful. It allows a vision system in an autonomous car to recognize pedestrians, lanes, signs, bicycles, trucks, and free space. It allows a robot to identify objects on a shelf. It allows predictive maintenance software to notice patterns before a bearing fails. It allows a recommender to infer interest, a medical system to flag risk, and a language model to produce fluent explanations. But prediction is not the same as control, and average behavior is not the same as safe behavior. A model trained to predict what usually happens has learned the center of past experience. A safety controller must ask what could happen at the edge, in the corner, under the rare combination of rain, glare, sensor fault, tired operator, missing sign, unusual body, malicious input, or unexpected human improvisation.

This is one of the deepest differences between much of AI and much of control engineering. Prediction often rewards being right on average. Control, especially safety-critical control, is haunted by the cases that do not occur often. An aircraft flight-control law cannot be designed only for the median gust. A nuclear reactor protection system cannot assume the operator will behave like the average operator under average pressure. A medical infusion pump cannot be safe only for the usual patient. A self-driving car cannot treat an unusual pedestrian as statistical noise simply because the training set contained more ordinary pedestrians. A lighthouse was valuable not because most nights were clear, but because some nights were black and wet and full of rock. "Our model is correct ninety-nine percent of the time," Daniel says in the control-room meeting. Ingrid answers, "Then tell me about the one percent." A manager says, "One percent is excellent." Ingrid turns to him. "One percent of what? If it is one percent of movie recommendations, perhaps. If it is one percent of landings, doses, braking events, or shutdowns, it is a graveyard with a nice dashboard."

Control engineers have formal ways of being pessimistic. They design for margins, disturbances, delays, uncertainties, actuator limits, sensor faults, saturation, noise, and worst cases. They ask whether a system remains stable when parameters change, whether a controller becomes aggressive when measurements are delayed, whether a safety interlock fails safe, whether redundancy is diverse enough that one hidden assumption cannot defeat every channel. They know that a beautiful response in simulation may become dangerous when the world supplies friction, wear, heat, corrosion, bad calibration, or people doing something locally reasonable and globally disastrous. AI engineers increasingly know these things too, but the culture of machine learning has often celebrated performance on test sets, benchmarks, leaderboards, and average loss. The two cultures need each other. AI can give control better eyes and richer predictions. Control can give AI a discipline of constraints, stability, verification, fallback, and humility. In a laboratory, imagine an AI researcher named Maya showing a perception model to a control engineer named Oskar. "It detects cyclists much better than the old system," she says. Oskar nods. "Good. What does it do when it is unsure?" "It outputs a probability." "What does the controller do with that probability?" "It depends." "Then that is where the safety argument begins, not where it ends."

The sensor itself has changed. In older control systems, a sensor often measured a physical quantity directly: temperature, pressure, speed, position, flow, voltage, current, level. Such sensors could be wrong, but their wrongness had a familiar flavor: drift, bias, noise, delay, failure, miscalibration. AI-based sensors can be stranger. A camera plus neural network may report pedestrian, lane marking, stop sign, tumor boundary, ripe fruit, damaged part, smoke plume, suspicious behavior, or empty room. The measurement is no longer merely a value; it is an interpretation. It may fail because the lens is dirty, because the lighting is unusual, because the object is rare, because the training data was biased, because an adversarial pattern confuses the classifier, because the world has changed, or because the category itself is ambiguous. When the thermostat mistakes a professor sitting still for an empty room, the error is comic. When an autonomous vehicle's perception system mistakes a person, a trailer, a shadow, or a construction zone, the error may become moral before it becomes mechanical. The controller acts on what the sensor says the world is. If the sensor is an AI interpretation, then part of the world has entered through a learned guess.

This does not mean AI should be kept out of control systems. It means AI must be placed inside architectures that understand uncertainty. A vision model may help a car see; it should not be treated as an oracle. A predictive model may help a grid operator anticipate congestion; it should not silently hide the assumptions behind its forecast. A medical model may help a doctor notice a pattern; it should not turn the patient into a statistical average wearing a name bracelet. A robot assistant may understand speech well enough to help Arne find his glasses; it should also know when to call Liv or a nurse instead of pretending competence. The human in the loop becomes especially important where meaning, responsibility, and exception meet. The human may not calculate faster than the model, but the human may know that the calculation is pointed at the wrong question. Ingrid says this to Daniel after the meeting while they stand beside a coffee machine that sounds like a small failing compressor. "I do not need the system to admire my experience," she says. "I need it to tell me what it sees, what it does not see, and how much time I have." Daniel asks, "And then?" "Then I can be useful instead of decorative."

Designing for a useful human is harder than leaving a human in the loop as a slogan. If automation handles normal operation, the human receives fewer chances to practice. Skill decays. Attention drifts. The operator becomes a monitor of boredom punctuated by terror. When the system suddenly asks for intervention, it may do so precisely because the situation is unusual, time-limited, and difficult: the worst moment to return control to someone who has been excluded from the preceding context. This problem appears in aircraft, cars, process plants, hospitals, and military systems. "Take over now" is not a plan if the person does not know what has happened, what the automation has tried, which sensors are trusted, what constraints are active, and what failure is unfolding. A pilot cannot be treated as a spare actuator. A doctor cannot be treated as a rubber stamp. A driver cannot be treated as an emergency backup after being invited to disengage. A content moderator cannot be expected to clean at human speed what an algorithm amplified at machine speed. The handoff is not a button. It is a relationship built before the emergency.

The earlier chapters give us several patterns. The governor worked because Tom remained close enough to understand the engine and suspicious enough to inspect the mechanism. The grid failed in part because operators did not receive the picture they needed when alarms and software broke down. Chernobyl teaches that being physically in a control room does not mean having real permission to stop. The Gripen crash reminds us that human intention must be translated through control laws whose behavior the pilot can learn and trust. Smart buildings show that occupants themselves may be sensors when the installed sensors miss discomfort. Robots show that autonomy is most humane when it removes needless burden without removing human relation. The feed chapter reverses the usual phrase: sometimes the human in the loop is not the supervisor but the controlled variable, nudged and measured by a system whose objective the user did not choose. These are not separate stories. They are one argument repeated at different scales. Automation does not remove humanity from the loop. It moves humanity around, sometimes upward into supervision, sometimes downward into data, sometimes outward into responsibility without power.

There is a dangerous fantasy that enough intelligence will eventually eliminate the need for human judgment. It is dangerous not because machines cannot become extraordinarily capable, but because judgment is not merely the filling of gaps left by inadequate computation. Judgment includes choosing goals, interpreting values, recognizing when the formal problem is not the real problem, accepting responsibility, and deciding what must not be optimized. A hospital can optimize bed turnover and still fail care. A feed can optimize engagement and still damage attention. A factory can optimize output and still destroy bodies or communities. A robot can optimize task completion and still make a person feel managed rather than helped. An AI system can produce the most likely answer and still be wrong in the way that matters. In an imagined ethics review, Maya says, "The model predicts what a good operator would do." Ingrid asks, "Which good operator? On which day? Under whose incentives?" Daniel says, "We trained on historical decisions." Ingrid answers, "Then you trained on our habits, our shortcuts, our blind spots, and our courage, all mixed together. Do you know which is which?"

The answer is not to romanticize humans. Humans are tired, biased, frightened, proud, distracted, hierarchical, and sometimes foolish. Human judgment caused disasters as surely as it prevented them. Operators disabled safety systems at Chernobyl. Drivers overtrust assistance systems. Doctors miss diagnoses. Pilots can become confused. Managers reward speed over safety. Users click what harms them. The human in the loop is not a magic ingredient sprinkled over automation to make it ethical. The human must be designed for as carefully as any sensor or actuator. That means training, authority, time, interface clarity, meaningful alarms, explainable system state, practiced handoffs, and the right to stop without punishment. It means avoiding alarm floods that teach people to ignore alarms, dashboards that hide uncertainty behind green icons, and procedures that exist only to protect institutions after failure. It means asking whether the person in the loop has information, competence, and power. Without all three, the loop is a theater.

Perhaps the best systems will be neither human-centered in a sentimental sense nor machine-centered in a technocratic sense, but responsibility-centered. They will ask what must be sensed, who must decide, what can be automated safely, what must remain interruptible, what should be forbidden, and how learning systems will be monitored after deployment. They will treat AI as a powerful component, not as a replacement for engineering discipline. They will use prediction, but not confuse prediction with permission. They will optimize, but only inside boundaries chosen through public, professional, and moral argument. They will keep humans close enough to understand and far enough from avoidable danger. They will allow machines to do what machines do well: repeat, measure, calculate, remember, warn, and act quickly. They will ask humans to do what humans must not abandon: choose ends, recognize exceptions, care about persons, and say no.

Near the end of Ingrid's meeting, after the managers have left and the engineers have gathered their laptops, Daniel stays behind. He is not offended anymore. He is tired in the useful way that follows a hard conversation. "So what do you want changed?" he asks. Ingrid looks at the wall of screens. "First, I want the system to show uncertainty, not just recommendations. Second, I want it to explain which sensors drove the recommendation. Third, I want simulation drills where operators fight the automation and the automation is allowed to be wrong. Fourth, I want a mode where it slows down before irreversible actions unless the plant is already in immediate danger. Fifth, I want management to sign a sentence saying an operator may stop the process without career suicide." Daniel types. "That last one is not software." "Most safety is not software." He smiles despite himself. "You know, for someone suspicious of automation, you have a detailed feature list." Ingrid picks up her coat. "I am not suspicious of automation. I am suspicious of loneliness." "Loneliness?" "A human alone with a machine that everyone else trusts too much. A machine alone with a goal no human has questioned. A user alone inside a feed. An old man alone with a care robot because visits became inefficient. A pilot alone inside a control law he cannot feel. A grid operator alone without alarms. Bad loops are lonely loops."

That may be the chapter's lesson. The human in the loop is not there because humans are always wiser than machines, faster than machines, or more reliable than machines. Often we are not. The human is there because control finally answers to purposes that cannot be found inside the error signal. Safety is not only stability. Comfort is not only temperature. Care is not only task completion. Navigation is not only position. Driving is not only motion. A feed is not only engagement. A factory is not only throughput. A society is not only a dashboard. The loop needs people not as decorations, scapegoats, or emergency spare parts, but as responsible participants with real information and real authority. Machines can correct motion, pressure, speed, light, dose, trajectory, and recommendation. They can even help us see patterns we would miss. But when the system asks what should be protected, what risk is acceptable, what failure would be unforgivable, and whose life is being optimized, the answer must come from beyond the controller. It must come from humans willing to remain in the loop not only to command, but to care.